International Conference on Electromagnetic Fields, Signals and BioMedical Engineering 2024

IOP Conference Series: Materials Science and Engineering
Volume 1320

Cluj-Napoca, Romania
6-8 June 2024

ISBN: 979-8-3313-1006-6
ISSN: 1757-8981

Printed from e-media with permission by:

Curran Associates, Inc.
57 Morehouse Lane
Red Hook, NY 12571

Institute of Physics
Dirac House, Temple Back
Bristol BS1 6BE UK

Phone: 44 1 17 929 7481
Fax: 44 1 17 920 0979

techtracking@iop.org

Additional copies of this publication are available from:

Curran Associates, Inc.
57 Morehouse Lane
Red Hook, NY 12571 USA
Phone: 845-758-0400
Fax: 845-758-2633
Email: curran@proceedings.com
Web: www.proceedings.com

TABLE OF CONTENTS

Between June 6 and 7, 2024, the Technical University of Cluj-Napoca hosted the second edition of the scientific conference ICEMS-BIOMED - International Conference on Electromagnetic Fields, Signals and BioMedical Engineering - dedicated to the research in electrical and biomedical engineering.

Organized under the aegis of EUT+ and based on the collaborative relations between prestigious academic institutions from Romania (Technical University of Cluj-Napoca, „Nicolae Bălcescu" Land Forces Academy of Sibiu and „Ştefan cel Mare" University of Suceava), the 2024 edition enjoyed the presence of researchers from 8 European countries: Spain, Greece, Poland, Latvia, Austria, Switzerland, Czech Republic and Romania.

As in the case of the previous edition, the conference was held in a hybrid format; following the evaluation process, out of a total of 40 papers initially uploaded on the site, only 27 were presented, the 13 rejected papers representing a percentage of 32.5%.

The Conference was attended by 4 invited speakers, who held lectures of forty minutes each, the rest of the papers being presented exclusively orally. In this way, a total of 27 papers were presented: 4 by invited speakers and 23 regular papers.

The main objective of this scientific event was to create emulation in the academic world in topics of great relevance and interest and to offer an opportunity to young researchers to present their newest results in approaching the following topics:

- Electromagnetic Fields
- Electromagnetic Compatibility
- Interactions between Electromagnetic Waves and Biological Systems
- Biomedical Engineering
- Modern Solutions for Diagnostics and Treatment

The event took place in the new conference center of the Technical University of Cluj-Napoca (UTC-N Hub) and in the presentation room there were about 50 people: authors (senior researchers or doctoral students), master's students, sponsors and guests. The regular presentations took place over a period of 15 minutes - 10 minutes for the presentation itself and 5 minutes for discussions and questions and the online presentations were made through the Teams platform.

This scientific event was made possible thanks to the efforts of generous people, who are grouped into the following committees:

Steering committee

Prof. univ.dr.ing. Vasile ȚOPA
Gl. Bg. Prof. univ.dr.ing. Ghiță BÂRSAN
Prof. univ. dr. Mihai DIMIAN
Prof. univ.dr.ing. Radu MUNTEANU
Prof. univ.dr.ing. Adrian GRAUR

Conference Co-Chairs

1. **Electromagnetic fields**
   - Jolanta Karpowicz, Central Institute for Labour Protection – National Research Institute, Poland
   - Simona Miclaus, „Nicolae Bălcescu" Land Forces Academy, Sibiu, Romania
2. **Electromagnetic compatibility**
   - Călin Munteanu, Technical University of Cluj-Napoca, Romania
   - Paul Bechet, „Nicolae Bălcescu" Land Forces Academy, Sibiu Romania
3. **Interactions between Electromagnetic Waves and Biological Systems**
   - Laurențiu Dan Milici, „Ştefan cel Mare" University, Suceava, Romania
   - Radu Ciorap, „Gr. T. Popa" University of Medicine and Pharmacy, Iași, Romania
4. **Biomedical Enginneering**
   - Mihai Munteanu, Technical University of Cluj-Napoca, Romania
   - Dan Rafiroiu, Technical University of Cluj-Napoca, Romania
5. **Modern Solutions for Diagnostics and treatment**
   - Dan-Sebastian Dîrzu, Cluj County Clinical Hospital
   - Octavian Andercou – „Iuliu Hațieganu" University of Medicine and Pharmacy, Cluj-Napoca, Romania

Scientific and Reviewers Committee

1. Joaquin Roca Gonzalez, Polytechnic University of Cartagena UPCT, Spain
2. Jurijs Dehtjars, Riga Technical University RTU, Latvia
3. Diana Porţan, Patras University, Greece
4. Jolanta Karpowicz, Central Institute for Labour Protection – National Research Institute, Poland Warszawa, Poland
5. Dan Mândru, Technical University of Cluj-Napoca, Romania
6. Doina Pîslă, Technical University of Cluj-Napoca, Romania
7. Mihai-Stelian Munteanu, Technical University of Cluj-Napoca, Romania
8. Dan Rafiroiu, Technical University of Cluj-Napoca, Romania
9. Alexandru Cristian Lodin, Technical University of Cluj-Napoca, Romania
10. Adina Giurgiuman, Technical University of Cluj-Napoca, Romania
11. Claudia Alana Constantinescu, Technical University of Cluj-Napoca, Romania
12. Ioan Marius Purcar, Technical University of Cluj-Napoca, Romania
13. Sergiu Iulian Andreica, Technical University of Cluj-Napoca, Romania
14. Adrian Bojiţă, Technical University of Cluj-Napoca, Romania
15. Anca-Iulia Nicu, Technical University of Cluj-Napoca, Romania
16. Angela Lungu, Technical University of Cluj-Napoca, Romania
17. Simona Vlad, Technical University of Cluj-Napoca, Romania
18. Cătălin Popa, Technical University of Cluj-Napoca, Romania
19. Ioan Valentin Sita, Technical University of Cluj-Napoca, Romania
20. Robert Groza, Technical University of Cluj-Napoca, Romania
21. Paul Bechet, „Nicolae Bălcescu" Land Forces Academy of Sibiu, Romania
22. Simona Miclăuș, „Nicolae Bălcescu" Land Forces Academy of Sibiu, Romania
23. Emil Şorecău, „Nicolae Bălcescu" Land Forces Academy of Sibiu, Romania
24. Mirela Şorecău, „Nicolae Bălcescu" Land Forces Academy of Sibiu, Romania
25. Aldo de Sabata, „Politehnica" University Timişoara, Romania
26. Radu Ciorap, „Gr. T. Popa" University of Medicine and Pharmacy of Iași, Romania
27. Călin Corciova, „Gr. T. Popa" University of Medicine and Pharmacy of Iași,

IOP Conf. Series: Materials Science and Engineering **1320** (2024) 011001 doi:10.1088/1757-899X/1320/1/011001

Romania

28. Dan Milici, „Ştefan cel Mare" University of Suceava, Romania
29. Oana Geman, „Ştefan cel Mare" University of Suceava, Romania

Organizing Committee

Mihai-Stelian Munteanu, Technical University of Cluj-Napoca, Romania

Claudia Păcurar, Technical University of Cluj-Napoca, Romania

Adina Giurgiuman, Technical University of Cluj-Napoca, Romania

Claudia Constantinescu, Technical University of Cluj-Napoca, Romania

Sergiu Iulian Andreica, Technical University of Cluj-Napoca, Romania

Paul Bechet, „Nicolae Bălcescu" Land Forces Academy of Sibiu, Romania

Simona Miclăuş, „Nicolae Bălcescu" Land Forces Academy of Sibiu, Romania

Robert Helbet, „Nicolae Bălcescu" Land Forces Academy of Sibiu, Romania

Laurenţiu Dan Milici, „Ştefan cel Mare" University of Suceava, Romania

Radu Ciorap, „Gr. T. Popa" University of Medicine and Pharmacy of Iaşi, Romania

Local Organizing Committee, from the Technical University of Cluj-Napoca

Mihai Munteanu

Claudia Răcăşan Păcurar

Adina Giurgiuman

Claudia Constantinescu

Sergiu Iulian Andreica

Adrian Bojiţă

Alexandru Lodin

Marian Răzvan Gliga

SPONSORS ACKNOWLEDGEMENTS:

EnergoBit **TEMCO** **KARL STORZ** **Rohde&Schwarz**

Signatories of the Preface,

EDITORS OF THE PROCEEDINGS VOLUME:

- Mihai Munteanu, Technical University of Cluj-Napoca, Romania
- Paul Bechet, „Nicolae Bălcescu" Land Forces Academy, Sibiu, Romania
- Simona Miclăuș, „Nicolae Bălcescu" Land Forces Academy, Sibiu, Romania
- Laurențiu Dan Milici, „Ştefan cel Mare" University, Suceava, Romania
- Radu Ciorap, „Gr. T. Popa" University of Medicine and Pharmacy, Iași, Romania

IOP Conf. Series: Materials Science and Engineering **1320** (2024) 011002 doi:10.1088/1757-899X/1320/1/011002

Peer Review Statement

All papers published in this volume have been reviewed through processes administered by the Editors. Reviews were conducted by expert referees to the professional and scientific standards expected of a proceedings journal published by IOP Publishing.

1. Type of peer review: Double anonymous
2. Conference submission management system: Morressier
3. Number of submissions received: 40
4. Number of submissions sent for review: 32
5. Number of submissions accepted: 24
6. Acceptance Rate (Submissions Accepted / Submissions Received × 100): 60 %
7. Average number of reviews per paper: 2.00
8. Total number of reviewers involved: 21
9. Contact person for queries:
   Name: Mihai MUNTEANU
   Affiliation: Technical University of Cluj-Napoca
   Email: Mihai.Munteanu@ethm.utcluj.ro

IOP Conf. Series: Materials Science and Engineering **1320** (2024) 012001 doi:10.1088/1757-899X/1320/1/012001

5G FR1 Exposure Measurements: Comparison of Different IEC 62232:2022 Extrapolation Methods

M. Ruttner[1], H. Keller[2], B. Strutz[1], D. Šuka[3], S. Qahtan[4]

[1] A1 Telekom Austria AG, Vienna, Austria;
[2] Narda Safety Test Solutions GmbH, Pfullingen, Germany;
[3] Faculty of Electrical Engineering, University of East Sarajevo, East Sarajevo, Bosnia and Herzegovina;
[4] Faculty of Engineering, University Putra Malaysia, 43400 UPM Serdang, Malaysia

E-mail: manfred.ruttner@a1.at

Abstract. Currently, mMIMO (massive Multiple Input Multiple Output) antennas with beamforming are primarily used in 5G NR (New Radio) in the sub-6 GHz range at 3.5 GHz (Frequency Range 1, FR1). The assessment of the theoretical or actual maximum of human exposure to RF field strength or power density according to international standard IEC 62232:2022 is significantly more difficult at base stations (BSs) with beamforming technology than with conventional antenna technologies. This study compares different measurement and exposure evaluation methods from IEC 62232:2022 with in-situ measurements. A case study, developed for the IEC TR 62669 Ed. 3 originally included an analysis of exposure levels at 4 measurement points (A, B, C and D) in the surroundings of three 5G NR BSs, whereas this study includes analysis from both the original and one additional measurement point (E).

1. Introduction

5G New Radio (NR) network evolves based on frequency-resource availability, financial metrics, and speed to provide services for a large population. A global mobile amount of traffic, according to [1], has been expected to increase more than 650 times considering the 20-year period (from 7.462 EB/mo in 2010 to approximately 5016 EB/mo in 2030). Due to the exponential growth of mobile subscribers, multimedia applications, and data applications, there is an increasing demand for wireless systems [2], which effectively illustrates the significance of the development and evolution of mobile communication technology. Fifth-generation (5G) technology promises to meet the needs that have resulted from the continually increasing expectations of mobile network subscribers, along with those imposed by innovative ideas like connected and self-driving vehicles, advanced manufacturing, or smart cities.

5G NR mobile communication is currently being implemented worldwide and is anticipated to achieve the intended thousandfold increase in network capacity with new and more advanced capabilities [1]. Based on the Ericsson Mobility Report published in June 2020 [3], there should be over 2.8 billion 5G subscriptions worldwide by the end of 2025. The report predicts that roughly half of all mobile data traffic will be carried by 5G networks [3]. Standalone (SA) mode provides all the benefits and full features of 5G NR (enhanced Mobile Broadband (eMBB), massive Machine Type Communication (mMTC), and Ultra-Reliable Low Latency Communications (URLCC)).

However, the introduction of the 3,5 GHz frequency range leads to the need for additional antennas, i.e. mMIMO antennas, to be installed at existing or new locations, especially in densely populated areas. As a consequence, the estimation of human exposure to electromagnetic fields (EMFs) from 5G systems is not a trivial task and needs to be carefully analyzed. Therefore, it is crucial to check compliance of such fields with exposure limits prescribed by international guidelines from the International Commission on Non-Ionizing Radiation Protection (ICNIRP) [4], Institute of Electrical and Electronics Engineers (IEEE) [5], Federal Communications Commission (FCC) [6] or some stricter national guidelines.

To control EM emissions from BSs in general, the measurement practice, defined in technical standards [7, 8], uses measurement methods based on frequency-selective and code-selective options. The former measures the current exposure in one frequency band over a defined period, whereas the latter measures and decodes available signals on air and gives the possibility to extrapolate those signals to the actual or theoretical maximum exposure. No matter which approach is being used for the EMF exposure evaluation, the BS signal contains a constant part (reference or pilot signals) whose level is nominally independent of the network traffic load. This signal is suitable for the development of extrapolation methods. Crucial importance in EMF exposure assessments based on extrapolation methods is the stability of reference signals, as discussed in [9, 10, 11, 12, 13, 14].

The rest of the paper structure is as follows. Section II describes the background and importance of extrapolation methods in human exposure assessments. Section III gives an overview of some BS operation parameters and details on MPs. Section IV compares the extrapolated results of the actual maximum exposure delivered by extrapolation methods described in B.8.5 [7] and Annex E [7]. The results and related discussion are depicted in Section V. Finally, the last section concludes the paper highlighting directions for future work and analysis.

2. Background of extrapolation methods

It is well known that the field strength at mobile radio communication systems varies over time. Besides fading effects, this mainly happens due to the power control of the mobile radio systems, i.e., the fact that only as much high-frequency power is transmitted as is necessary for data transmission. On the other side, a constant signal is required for a mobile device to detect a mobile radio system and register in the network. However, such a signal only accounts for a small part of the maximum available transmission power but can be used to assess the human exposure at MP.

To speed up the measurement process and make it more feasible for technicians involved in the measurements on the field, recent practices to assess human exposure to 5G signals use approaches for determining the maximum value of the electric field, and in particular, the theoretical maximum and the actual maximum ones [7].

As mentioned above, the power of the pilot signal is constant whereas the transmitted power of BS depends on the current traffic load and with antennas with beamforming the exposure has an additional variation due to the spatial and temporal distribution of individual beams. Therefore, by measuring the field strength of the pilot signal it is possible to calculate the maximum exposure from BS under the highest traffic load. The actual maximum approach is necessary because of the beamforming of current mMIMO antennas. It is the time-averaged exposure that takes into account the variation in traffic load and beamforming and gives a realistic value for the exposure. On the other hand the 'theoretical maximum' is rarely reached during the BS operation.

Beamforming is a technology that allows the focusing of the signal in a specific direction.

Figure 1. Measurement point A: BS antennas are in an orange circle. Distance to BS antennas is 428 m.

Conventional antennas are broadcasting in all directions within their transmission range. Beamforming is achieved by adjusting the phase and amplitude of the signal at each antenna element in the array. By doing so, the signal strength and quality can be improved for the intended receiver, leading to better performance and increased capacity in the wireless network.

Finally, the code-selective method and analogue measurement of the SSB assumes the stochastic nature of the transmitted signals from the BS considering the real achievable maximum EM emissions, antenna radiation pattern, use of beamforming in mMIMO systems, flexible numerologies, and Time Division Duplexing (TDD) schemes representing that way an evaluation method suitable for future development of 5G networks [7, 15].

3. Description of evaluation sites

Measurements for the original case study conducted for [8] were performed at four MPs considering three different 5G NR BSs operating in FR1. One site included two MPs, whereas the rest included only one. All MPs have been selected a priori. This case study was also published as an Application Note [16] for the selected measurement equipment Narda SRM-3006 (Spectrum Analyzer, SA). However, for this study, a fifth MP was analysed to re-evaluate the results of the original case study with an MP near a mobile communications system in an urban environment. It should be noted that the fifth MP is not accessible to the general public (the roof terrace is only accessible to authorized personnel). For all five MPs, a three-axis isotropic antenna for the frequency range 420 MHz - 6 GHz was used. Neither spatial averaging nor spatial maximum search was applied. The MPs were selected with the objective that only one sector of the examined BS dominates the expected exposure. Figures 1 to 5 illustrates the positions of MPs.

Technical characteristics of 5G NR BSs were:

- The central frequency of the 5G NR signal was 3,54 GHz;
- The subcarrier spacing and bandwidth, respectively, were 30 kHz and 100 MHz;
- The BS operated in TDD mode with 70 % downlink time;
- Each sector antenna of the BS had 64 ports;

Figure 2. Measurement point B: Distance to BS antennas is 58 m.

Figure 3. Measurement point C: BS antennas are in an orange circle. Distance to BS antennas is 265 m.

- The configured maximum power per port was 5 W;
- The maximum gain of the traffic beams was 23,3 dBi for MP A-D and 24,8 dBi for MP E;
- For the broadcast signals, four beams were configured for MPs A-D and six beams for MP E;
- Maximum gain of broadcast beams was 21 dBi;
- There was no power control activated during the operation of BSs;
- There was no boosting of resource elements (REs) of broadcast signals activated;
- The BS maximum configured transmitted power is 320 W;
- Four factors used to calculate the extrapolation factor $F_{extBeam}$ according to B.8.5 [7] are given in Table 1.

IOP Conf. Series: Materials Science and Engineering **1320** (2024) 012001 doi:10.1088/1757-899X/1320/1/012001

Figure 4. Measurement point D: Distance to BS antennas is 282 m.

Figure 5. Measurement point E: Distance to BS antennas is 37 m.

Table 1. Factors used to calculate the extrapolation factor $F_{extBeam}$.

Factor	Number	Remark
F_{TDC}	0,7	i.e. 70 %
F_{PR}	0,25	according to the CDF from similar sites and literature[1]
F_{PR}	3276	according to table E.11 [7]
F_B	1	no boosting of broadcast signals

[1] CDF: Cumulative Distribution Function Method is described in Section 6.2.3.2 of IEC 62232 [7].

The extrapolation factor $F_{extBeam}$ according to B.8.5 [7] depends on the directions from the relevant sectors to the MPs and was evaluated in detail. The elevation angles were calculated by the distance and height differences between the centre of the relevant sector antenna and the MPs. The distance and azimuth angles were derived from a Google map and were checked with binoculars with laser range finder and a compass. The vertical and horizontal attenuations were added to obtain the angle-dependent attenuations because no data was available for the combination of both angles.

4. Evaluation process

The purpose of this study was to compare the extrapolated results of the actual maximum exposure delivered by the following methods described in B.8.5 [7] and Annex E [7]:

- Calculation assuming free space environment;
- Measurement under forced maximum traffic load;
- Extrapolation based on decoded SSS measurements;
- Extrapolation based on analogue SSB measurements;
- Extrapolation based on the traffic time slots in the SSB frequency range.

4.1. Actual maximum exposure estimation using free space calculation

The actual maximum exposure is calculated directly using equation for free space calculation described in B.2.2 [7] and transmitted power, gain and distance parameters. The results are given in Table 2.

Table 2. Actual maximum exposure using free space evaluation.

MP	A	B	C	D	E
Description	suburban	at a car park	moderat urban	rural	urban
Distance [m]	428	58	265	282	37
Traffic Gain [dB]	10,63	-3,11	18,28	21,43	15,31
Calculated power density [Wm^{-2}]	0,000281	0,000647	0,004271	0,007789	0,112855

4.2. Frequency-selective measurement with forced maximum traffic load

These measurements were performed according to B.4.2.5.4.2 [7]. Maximum traffic was simulated with the nPerf application on an unshaped profile over a download duration of 60 s. The mobile phone that forced the traffic load was located at a distance of more than 1,5 m from the measuring antenna according to item g) in B.4.2.5.4.2 [7]. The mobile phone was checked to be connected to the expected Cell ID (PCI, Physical Cell Identity) and beam. It was also checked by measurements that the uplink had no relevant influence on the measurement results. The RMS trace of the SA was averaged over 30 seconds and integrated from 3.490 MHz to 3.590 MHz. After multiplication with $F_{PR} = 0{,}25$, from Table 1, the estimation of the actual maximum power density at the MPs was obtained. The results are given in Table 3.

Table 3. Actual maximum power density with maximum traffic load

MP	A	B	C	D	E
Maximum power density [Wm^{-2}]	0,00654	0,01033	0,03422	0,02753	0,8862
Actual maximum power density [Wm^{-2}]	0,001635	0,002583	0,008555	0,006883	0,22155

4.3. Extrapolation based on SSS decoding

These measurements were performed according to E.8.2.1.2 [7]. Only the strongest beam of the relevant sector was evaluated. The other beams and sectors were negligible. About eight isotropic decoder runs were performed. The maximum value of all 8 runs was selected as the measurement result. The results are given in Table 4.

Table 4. Actual maximum power density based on SSS decoding

MP	A	B	C	D	E
Power density per RE [μWm^{-2}]	0,9426	1,976	5,755	2,203	23,92
$F_{extBeam}$	1,36	1,55	1,01	6,49	21,18
Actual maximum power density [Wm^{-2}]	0,000736	0,001759	0,00333	0,008192	0,290448

4.4. Extrapolation based on analogue measurement of the SSB

These measurements were performed according to E.8.2.1.3.2 [7]. Since the zero-span mode of the SA is used, the readings of the isotropic three-axis measurement antenna were obtained by three separate measurements. The measurement bandwidth was 6,4 MHz. The result per RE was obtained by multiplying the measured power density by 30 kHz/6,4 MHz, to follow the item f) in E.8.2.1.3.1. [7]. The results are given in Table 5.

Table 5. Actual maximum power density based on analogue SSB measurement

MP	A	B	C	D	E
Power density on x-axis in 6,4 MHz BW [μWm^{-2}]	91,61	344,7	976,4	55,92	1291
Power density on y-axis in 6,4 MHz BW [μWm^{-2}]	51,78	85,83	83,78	157,0	2155
Power density on z-axis in 6,4 MHz BW [μWm^{-2}]	49,8	78,81	18,96	192,56	1774
Power density on total in 6,4 MHz BW [μWm^{-2}]	193,19	509,34	1079,1	405,48	5220
$F_{extBeam}$	1,36	1,55	1,01	6,49	21,18
Actual maximum power density [Wm^{-2}]	0,000707	0,002125	0,002927	0,007068	0,302311

Table 6. Actual maximum power density based on traffic time slots in the SSB frequency range

MP	A	B	C	D	E
Power density on x-axis in 6,4 MHz BW [μWm^{-2}]	395,1	420,8	5231	232,1	36690
Power density on y-axis in 6,4 MHz BW [μWm^{-2}]	296,4	449,7	2144	867,3	39840
Power density on z-axis in 6,4 MHz BW [μWm^{-2}]	242	958,2	1237	1147	3040
Power density on total in 6,4 MHz BW [μWm^{-2}]	933,5	1828,7	8612	2246,4	79570
Actual maximum power density [Wm^{-2}]	0,002509	0,004914	0,023143	0,006037	0,217574

4.5. Extrapolation based on the traffic time slots in the SSB frequency range

This method uses zero-span measurements. Instead of the SSB beams, time slots with maximum traffic load are evaluated. The measurements were taken during the forced maximum traffic load. A constant power density was expected in the downlink time slots which are not used for the SSB beams. The results are given in Table 6.

5. Results and Discussion

Table 7 shows the estimated actual maximum power density for the five MPs and five methods in absolute values. We used the forced maximum traffic load as the reference method because it was the most reliable and accurate method at the time of measurements.

Table 7. Comparison of the absolute actual maximum power density values

MP	A	B	C	D	E
Calculation under free space conditions [Wm^{-2}]	0,000281	0,000647	0,004271	0,007789	0,112855
Channel power during maximum traffic load [Wm^{-2}]	0,001635	0,002583	0,008555	0,006883	0,22155
Extrapolation based on SSS decoding [Wm^{-2}]	0,000736	0,001759	0,00333	0,008192	0,290448
Extrapolation based on analog SSB [Wm^{-2}]	0,000707	0,002125	0,002927	0,007068	0,302311
Extrapolation based on traffic time slots [Wm^{-2}]	0,002509	0,004914	0,023143	0,006037	0,217574

In future, with more active users, it might however be more difficult to attract all traffic in the direction of MP.

The free space calculation showed different results. At MPs A, B and E the results of the free space calculation showed lower results compared to the measurements. With this calculation method, no reflections are taken into account. The unavoidable reflections increase the average power density in a volume of a cube with an edge length of some wavelengths around the MP.

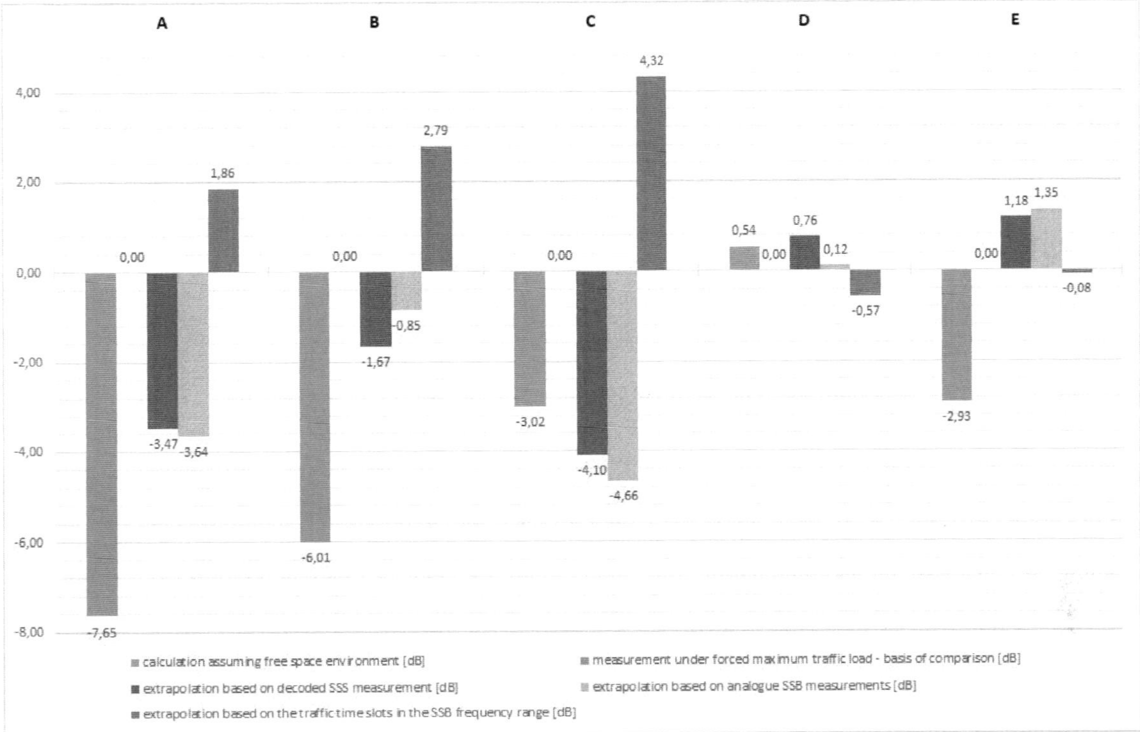

Figure 6. Comparison of the investigated methods with measurement under forced maximum traffic load as a basis

The amount of power density increase depends highly on the actual environment and the angle attenuation relative to boresight. At MPs with high angle attenuation, the expected power density increase due to reflections is high, too. In this case, reflectors are likely exposed to higher power densities compared to the MP itself. For MPs A, B, C and D, the horizontal angle attenuations, respectively, were 12,51, 5,65, 4,5 and 1,34 dB. Note that there is a good correlation between the horizontal angle attenuation and the measurement results.

Since the measurements were taken only at a single point without any spatial averaging, we expected additional uncertainty of several dBs in a reflective environment. Also, note that MP D was the only MP in a rural environment with no relevant reflectors. Here the calculation is very close to the measurement.

Since only two cuts of the antenna gain patterns were available, the gain into the actual direction was estimated by subtracting the horizontal and vertical attenuation from the boresight gain. This estimation is not the exact gain into the actual direction and might explain a part of the difference between the calculation assuming free space environment and the measurements. The uncertainty of the measurement device may explain a part of the difference, too.

The extrapolation based on the SSS decoding as well as analogue SSB measurement are very close to each other. It is worth mentioning that the analogue SSB measurement cannot distinguish between several cells while the SSS decoding can. In this campaign, this was not an issue because only one cell was relevant.

In the urban environment, we observed lower values than the reference (measurement under forced maximum traffic load, Figure 6). Possible reasons are that $F_{extBeam}$ is only correct under free space propagation conditions. In the rural environment, at MP D, the results are very close to the reference. The extrapolation based on the traffic time slots in the SSB frequency range showed higher values compared to the reference. The values would be even higher if the time slots with maximum power density would have been used. A possible reason is that the bandwidth extrapolation is only correct under free space propagation conditions. In the rural environment at MP D, the result was very close to the reference.

6. Conclusion

In this study, different measurement and evaluation methods taken from IEC 62232:2022 were compared at in-situ measurements at 5 MPs in the surroundings of 4G NR BS. A case study, developed for the IEC TR 62669 Ed. 3 originally included an analysis of exposure levels at 4 measurement points (A, B, C and D) in the surroundings of three 5G NR BSs, whereas this study includes analysis from both the original and one additional measurement point (E). The main findings of the study are:

- The different measurement methods and their resulting assessments of the actual maximum correlate relatively well;
- The extrapolation based on SSS decoding as well as on analogue SSB measurements are very close to each other;
- The extrapolation based on the traffic time slots in the SSB frequency range showed higher values compared to the reference except for MPs D and E;
- The free space calculation showed different results. At MPs A, B and E the results of the free space calculation show lower results compared to the measurements. With this calculation method, no reflections are taken into account. The more reflective objects can have an influence, the more inaccurate this calculation method becomes;
- In practice, determining the geometric relationship between the 5G NR BS antennas and MPs proves to be a major challenge. This factor, combined with the calculation based on angle attenuations in two planes only can lead to additional uncertainties;
- Considering the manufacturer's specified measurement uncertainty for used measurement system and measurement antenna in the corresponding frequency range from +2.0 dB to -2.6 dB, the differences appear to be acceptable in general;
- A very good agreement between all presented methods is evident in rural and low-reflective environments.

References

[1] Basharat S, Hassan S A, Pervaiz H, Mahmood A, Ding Z and Gidlund M 2021 *IEEE Wireless Communications* **28** 184–194
[2] Srivastava A, Gupta M S and Kaur G 2020 *Journal of Network and Computer Applications* **168** 102760
[3] Joshi P, Ghasemifard F, Colombi D and Törnevik C 2020 *IEEE Access* **8** 204068–204075
[4] 2020 Guidelines for limiting exposure to electromagnetic fields (100 kHz to 300 GHz) Tech. Rep. ICNIRP 2020 Health Physics
[5] IEEE 2019 *IEEE Std C95.1-2019 (Revision of IEEE Std C95.1-2005/Incorporates IEEE Std C95.1-2019/Cor 1-2019)* 1–312
[6] FCC 2020 Human exposure to radiofrequency electromagnetic fields and reassessment of FCC radiofrequency exposure limits and policies. A rule by the Federal Communications Commission published in the Federal Register on 04/01/2020
[7] IEC 2022 IEC Central Secretary, "IEC 62232:2022, Determination of RF field strength, power density and SAR in the vicinity of base stations for the purpose of evaluating human exposure Standard IEC 62232:2022 International Electrotechnical Commission Geneva, CH
[8] IEC 2019 IEC Central Secretary Project: IEC TR 62669 ED3 Case studies supporting IEC 62232 - Determination of RF field strength, power density and SAR in the vicinity of radiocommunication base stations for the purpose of evaluating human exposure, 106/623/CD Standard IEC TR 62669:2019 International Electrotechnical Commission Geneva, CH

[9] Betta G, Capriglione D, Cerro G, Miele G and D'Amata M S 2021 *IEEE Transactions on Instrumentation and Measurement* **70** 1–13
[10] Bernieri A, Betta G, Capriglione D, Cerro G, Miele G and Salone D'Amata M 2020 *IEEE Transactions on Instrumentation and Measurement* **70** 1–13
[11] Betta G, Cerro G, Miele G, Salone D'Amata M and Capriglione D 2020 *2020 IEEE International Instrumentation and Measurement Technology Conference (I2MTC)* (IEEE) pp 1–6
[12] Betta G, Capriglione D, Cerro G, Miele G, Migliore M D and Šuka D 2022 *IEEE International Instrumentation and Measurement Technology Conference (I2MTC), Ottawa, ON, Canada* 1–6
[13] Betta G, Capriglione D, Cerro G, Miele G, Migliore M D and Šuka D 2022 *2022 IEEE International Symposium on Measurements&Networking (M&N), Padua, Italy* 1–6
[14] Betta G, Capriglione D, Cerro G, Miele G, Wali S Q, Ruttner M, Sali A and Šuka D 2023 *2023 IEEE International Instrumentation and Measurement Technology Conference (I2MTC)* pp 01–06
[15] Colombi D, Joshi P, Xu B, Ghasemifard F, Narasaraju V and Törnevik C 2020 *Applied Sciences* **10** 5280
[16] Manfred Ruttner and Helmut Keller 2023 URL `https://www.narda-sts.com/index.php?eID=dumpFile&t=f&f=2559&dl=1&token=096f9d70e7d45e2c5080e4f08d4ca4ec9f7ed2fb`

IOP Conf. Series: Materials Science and Engineering **1320** (2024) 012002 doi:10.1088/1757-899X/1320/1/012002

A new and highly precise forecasting method based on multiple input Convolutional Neural Networks and Generative Adversarial Networks for training data set augmentation

Adrian-Valentin BOICEA[1]*and Adrian VULPE-GRIGORAȘI[2]

[1] Department of Electrical Power Systems, National University of Science and Technology Politehnica Bucharest, Bucharest, Romania
[2] NOVA LINCS, NOVA School of Science and Technology, NOVA University Lisbon, Portugal

*E-mail: valentin.boicea@upb.ro

Abstract. Forecasting is essential when it comes to power system operation. It is always equivalent with higher revenues and, implicitly, lower maintenance costs. That is why, it is the purpose of this work to introduce a new and highly precise wind power forecasting neural model based on a multiple input (multi-channel) Convolutional Neural Network (CNN) and a Generative Adversarial Network (GAN) for training data augmentation. The training of the CNN is conducted simultaneously with historical data related to both power generation and wind speeds.

1. Introduction

At present, forecasting the load or the power generation has become an essential part of the grid operation, given also the increased adoption of renewable sources, which complicate the things even more. That is because, on one hand, they are very intermittent in nature and, on the other, due to the fact that in various countries like, for example Romania, they tend to substitute altogether the central power system, especially when it comes to feeding the households. As such, the existing national power system is confronted with multiple and great difficulties since, originally, it was designed for a single direction of the power flow, from producer to the end customer. That is why, all the protection and metering devices have been designed for this paradigm and now they have to be changed at large scale in order to avoid the collapse of the grid in certain areas. Bearing that in mind, the high precision forecasting begins to play a cardinal role when planning the production capacities of the national power system from various conventional and renewable sources. Knowing the exact difference between the two, at national and distributed level, becomes equivalent with lower maintenance costs, more efficient operation and improved safety and security for the utility companies and final customers alike.

As a general rule, a 10 – 20 % improvement of the forecasting accuracy in supply chain could result in a 2-3 % increase in the revenues [1]. These 10-20 % represent in fact the usual difference in precision between the classical prediction models based on statistics and those that are based on Artificial Intelligence (AI). Sometimes, depending on the exact situation, the difference is even greater. This could be the case when the data in the time series is not too much affected by white noise or when the sample number is large enough to permit a highly accurate prediction. That, in turn, can also be used in detecting potential anomalies in the power generation and consumption. So, in the end, an accurate forecast means better planning and, of course, an increased profitability ratio for the company.

In order to rightly address the afore mentioned problem and to improve the forecasting accuracy, combined AI techniques have to be employed. Thus, in this work, a prediction model based on a multiple input (multi-channel) Convolutional Neural Network (CNN) and a Generative Adversarial Network (GAN) used to augment the training data set for the CNN will be presented[1]. Thus, one shall carry out a forecast of the generated electricity coming from a windfarm, based on historical timeseries containing data related to the power generation itself and to the wind speed. The end goal is to conduct a comparison between the prediction made with non-augmented data and the one carried out with augmented data.

The work is divided as follows: section 2 contains a short description of the installation from which the observations are coming, section 3 is devoted to the study of the neural networks and various metrics used to quantify the results, while section 4 is dedicated to conclusions and further research directions.

2. Short description of the windfarm

The power and weather data come from the Erie Shores windfarm at Port Burwell, Ontario, Canada and is available in RETScreen 4.1. According to [2], this facility comprises 66 General Electric SLE 1.5 wind turbines, possessing a maximum total capacity of 99 MW. The generated power is transmitted through 34.5 kV overhead lines to the Port Burwell substation. Here, the voltage is increased to 115 kV in order to be delivered to the Ontario transmission network. Thence, all the generated power coming from Erie Shores is sold to the Ontario Power Authority. Built in 2006, this facility represents approximately 4.8 % of the Ontario installed wind power capacity.

Both time series relating to the generated power and wind speed are sampled at 6 days, spanning a period from the 1 st. of June 2006 down to the 18 th. of February 2012. Thus, each of the training data categories contain 290 samples, making them suitable for this kind of analysis in which the forecast obtained with non-augmented data is compared to the forecast obtained with augmented data.

3. Model architecture and results

As stated previously, the main goal of this work is to observe the improvement in the forecast accuracy when carrying it out with an augmented training set. Hence, mention should be made of

[1] This concept based on multiple input CNN and GAN for training data set augmentation has been registered with the Romanian Intellectual Property Office (Oficiul Român pentru Drepturile de Autor), under the number 1386/29.03.2023. At the same time, the multiple input CNN is protected by a patent application registered with the Romanian State Office for Inventions and Trademarks, under the number RO135952A2. The research in this conference paper represents also the result of the following projects: Dezvoltarea Capacității Instituționale a Universității POLITEHNICA din București (DECIP), PFE ID 574, and EUBBC-Digital: Europe-Brazil-Bolivia-Cuba Capacity Building Using Globally Available Digital Learning Modules (EUBBC), 618925-EPP-1-2020-1-BR-EPPKA2-CBHE-JP – E+ CBHE.

the fact that the initial training data possess 290 samples (for both power generation values and wind speeds) whereas the augmented one will possess 580. In order to augment these two time series, a GAN will be employed, whose architecture can be seen in Figure 1. Both the GAN discriminator and the generator are represented by a CNN with one input, while the noise is Gaussian.

On the other hand, in order to obtain the forecast of the generated power as a function of the wind speed, a multiple input CNN will be used. Its architecture is depicted in Figure 2. This neural network will be employed for both cases, with non-augmented and with augmented training data sets.

The entire neural model has been implemented in Matlab 2024a.

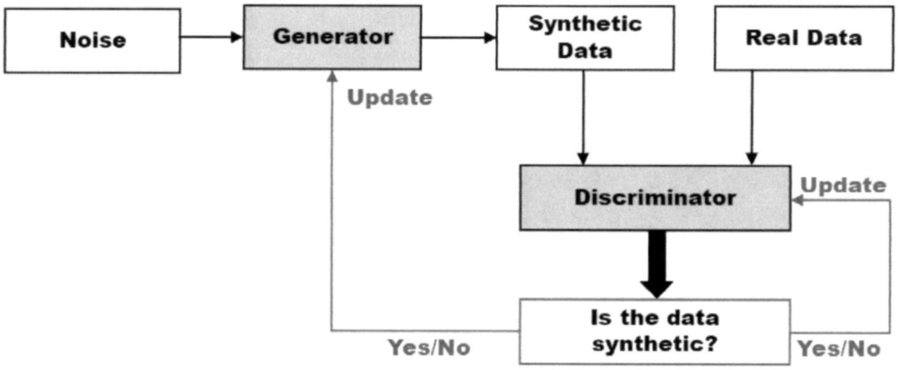

Figure 1. GAN architecture used for the time series augmentation in case of the power generation and wind speed.

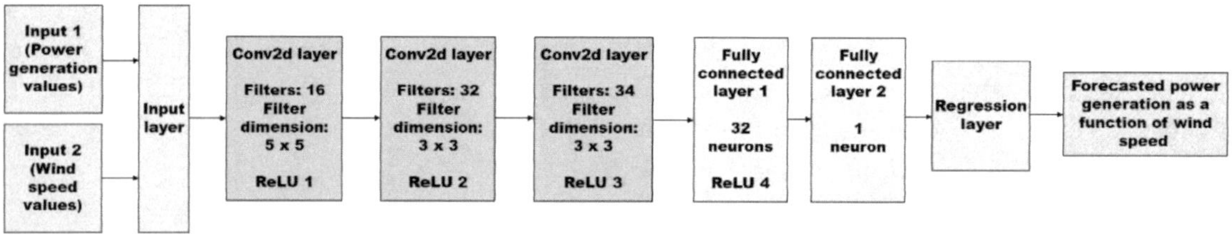

Figure 2. Multiple input CNN used for power generation forecast as a function of wind speeds.

One shall begin with the forecast based on the non-augmented training set.

The CNNs have been primarily dedicated to image processing. When it comes to time series, two consecutive values are related the same way as two consecutive pixels in an image. That is why these networks present optimal performances when used in forecasting. In this particular situation, the CNN must be trained to predict the power generation based, simultaneously, on historical data related to the generation itself and to the wind speeds. Thus, the CNN will have 2 inputs. This kind of approach is fully compatible to the concepts of image height and width. As it can be seen in Figure 2, input 1 is represented by the historical power generation values whereas input 2 is represented by the historical wind speeds. As such, input 1 will be assimilated to the image height and input 2 to the image width. This means that the filter will be simultaneously applied to both inputs and the network will be capable of learning more complex and relevant features than in the case of a single input. In order to test the robustness of the model, no data set has been normalized prior to ingestion, and one renounced to the batch normalization [3], as well (see Figure 2).

The applying of the filter on the two inputs is done using the convolution. The result will be a new image (that is, new values coming from the convolution operation between the filter and one of the two initial time series) which takes [4] place according to Equation (1):

$$R(i,j) = \sum_m \sum_n A(m,n)B(i-m+1, j-n+1) \tag{1}$$

where:

$R(i,j)$ –the end result of the bidimensional convolution;

$A(m,n)$ –the convolutional filter which is considered a matrix having m rows and n columns;

$B(i-m+1, j-n+1)$ –one of the two inputs which is considered a matrix having *i-m+1* rows and *j-n+1* columns.

In reality, A is the convolutional filter and B is one of the univariate time series pertaining either to the historical generated power or to the wind speeds. In doing so, each element of the filter is weighted combined with the corresponding values of the two timeseries and the aggregated result represents the new data possessing new and important characteristics like time and spatial patterns, trends and so on that, after a certain number of training epochs, will represent the actual forecast.

In Figure 3, one can see the initial variation of the generated wind power, without augmentation.

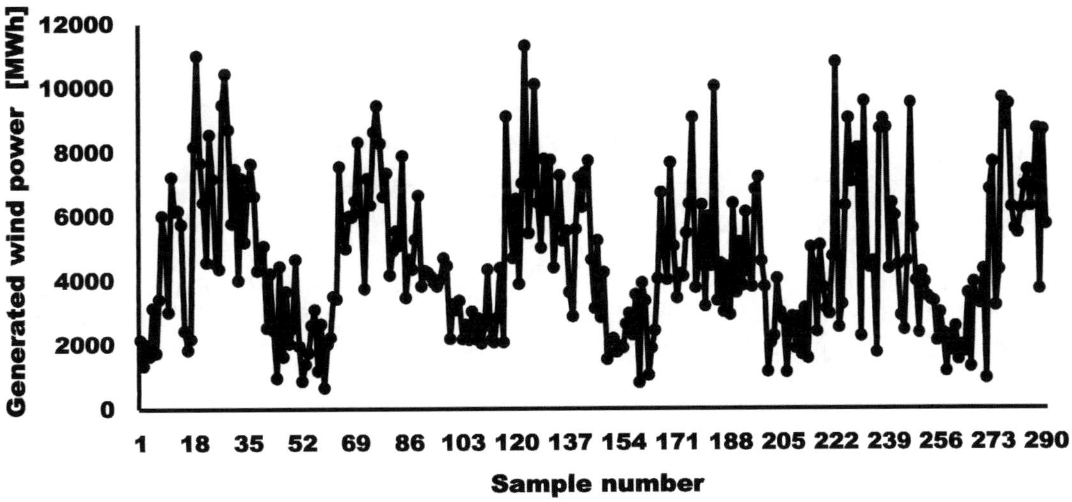

Figure 3. Non-augmented generated wind power between 01.08.2006-18.02.2012.

The Augmented Dickey-Fuller Test revealed that the time series in Figure 3 is stationary. That means that its mean, variance, and autocorrelation do not change over time.

In order to carry out the forecast, the data above has been split in 80 % training and 20 % validation data. The same test revealed once more that the training data is stationary whereas the validation data is not. The network was trained 200 epochs and its training RMSE (Root Mean Square Error) minimization is presented in Figure 4.

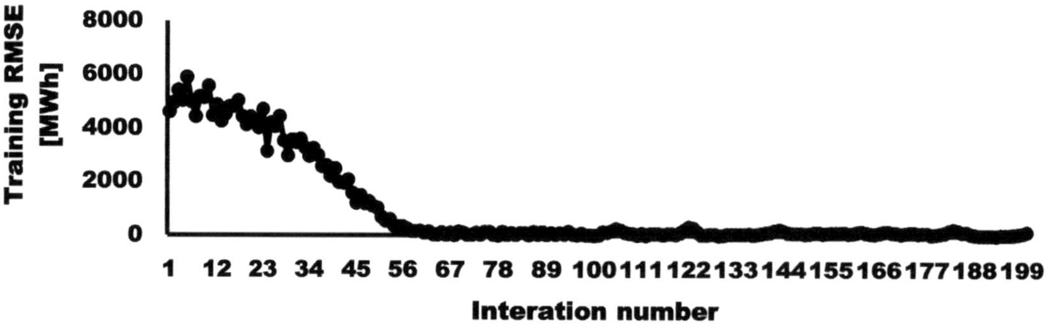

Figure 4. Minimization of the training RMSE.

In this case, the model will be validated on the last 58 samples, covering the period between 03.09.2007-18.02.2012. The result of the forecast can be seen in Figure 5.

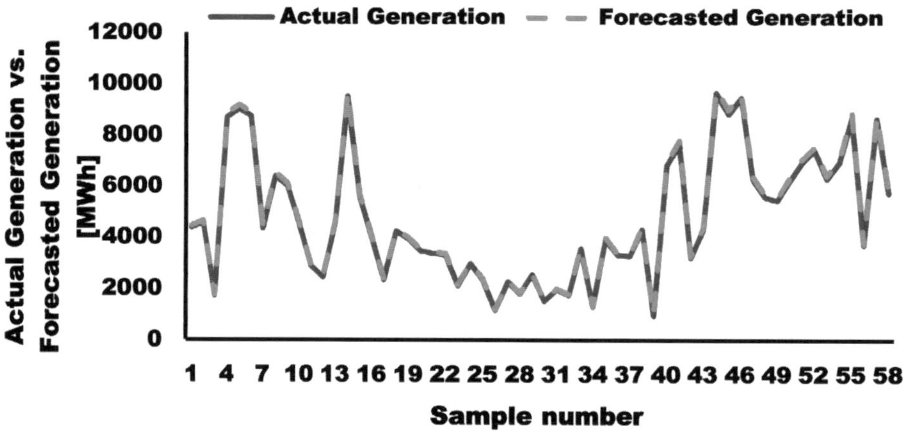

Figure 5. Actual generation vs. forecasted generation with non-augmented data.

The metrics used to assess the performance of the model are: the Mean Absolute Error (MAE), the Mean Bias Error (MBE), the Mean Absolute Percentage Error (MAPE), and the RMSE. Their formulas can be seen in the Equations (2), (3), (4), and (5).

$$MAE = \frac{1}{n} \cdot \sum_{i=1}^{n} |\hat{y}_i - y_i| \tag{2}$$

$$MBE = \frac{1}{n} \cdot \sum_{i=1}^{n} \hat{y}_i - \frac{1}{n} \cdot \sum_{i=1}^{n} y_i \tag{3}$$

$$MAPE = \frac{1}{n} \cdot \sum_{i=1}^{n} \frac{|y_i - \hat{y}_i|}{y_i} \cdot 100 \tag{4}$$

$$RMSE = \sqrt{\frac{1}{n} \cdot \sum_{i=1}^{n} (\hat{y}_i - y_i)^2} \tag{5}$$

where:
$\hat{y}_i$ –the forecasted value
y_i –the actual value
n – the number of samples.

In this case, the MAE= 89.7001 MWh, MBE=89.7001 MWh, MAPE=1.8849 %, RMSE=100.6 MWh.

The second part of this section is dedicated to the same type of forecast, using the same CNN and training parameters, but with two augmented training data sets. As mentioned above, this has been carried out with the GAN. Its training session for the generator, related to the augmentation of the electrical data, can be seen in Figure 6. That for the discriminator is very similar. The training for wind data augmentation presents the same type of variations, as well. Both the generator and the discriminator are represented by single input CNNs.

In order to assess the quality of the new synthetic time series, the Pearson, Kendall and, Spearman correlation coefficients between the GAN data and the initial data have been computed.

The Pearson correlation gives a measure of their linear dependence while the Kendall correlation evaluates to which extent the relative order of the values in a time series is similar or different to the relative order of the values in another set of observations. It is useful above all when the values are not normally distributed or when among them are outliers. This coefficient varies between [-1; 1].

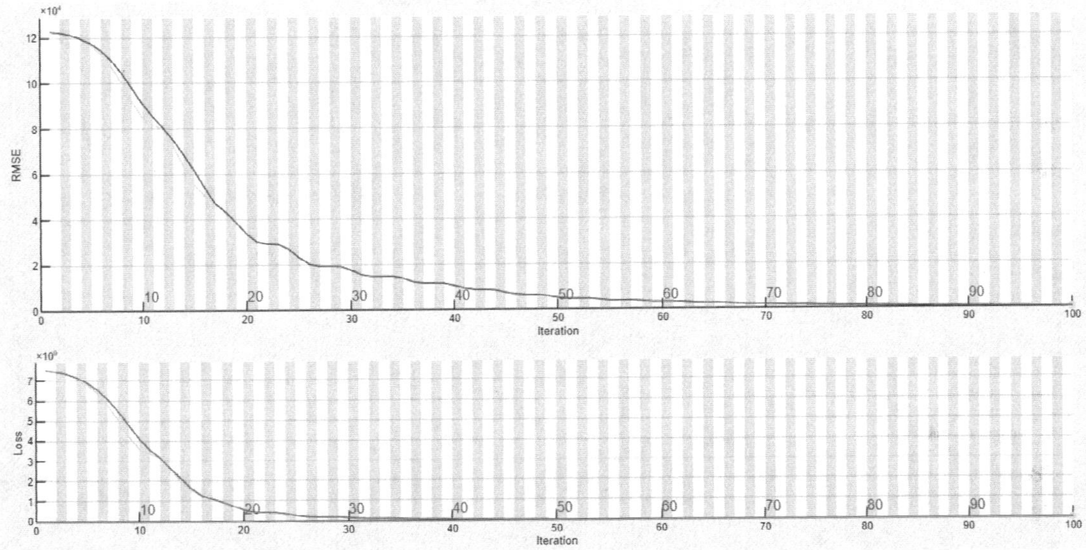

Figure 6. GAN generator training for electrical data augmentation.

-1 denotes a total mismatch between the two sets: when the values in one of them increase, the values in the other one decrease and vice versa. 1 means a total concordance between the sets while 0 is equivalent with no relationship between the sets.

The Spearman correlation coefficient takes into account the rankings of each observation in each of the two sets. If the ranks are perfectly correlated, this coefficient equals 1, if they are not correlated the coefficient equals -1, whereas 0 denotes a total lack of relation between the sets.

In this particular case, measured between the generator and the initial data, the Pearson, Kendall, and Spearman coefficients are 0.99992, 0.9915, and 0.9999, respectively.

The augmented power generation time series, containing 580 samples (the double of the initial size), is depicted in Figure 7. The increased number of sample cover the same initial period between 01.08.2006-18.02.2012.

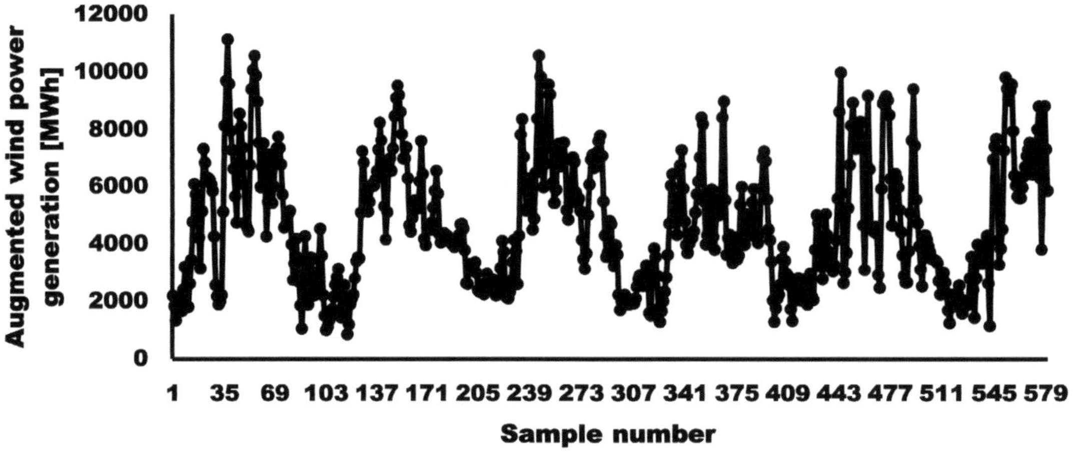

Figure 7. GAN augmented wind power generation for the period 01.08.2006-18.02.2012.

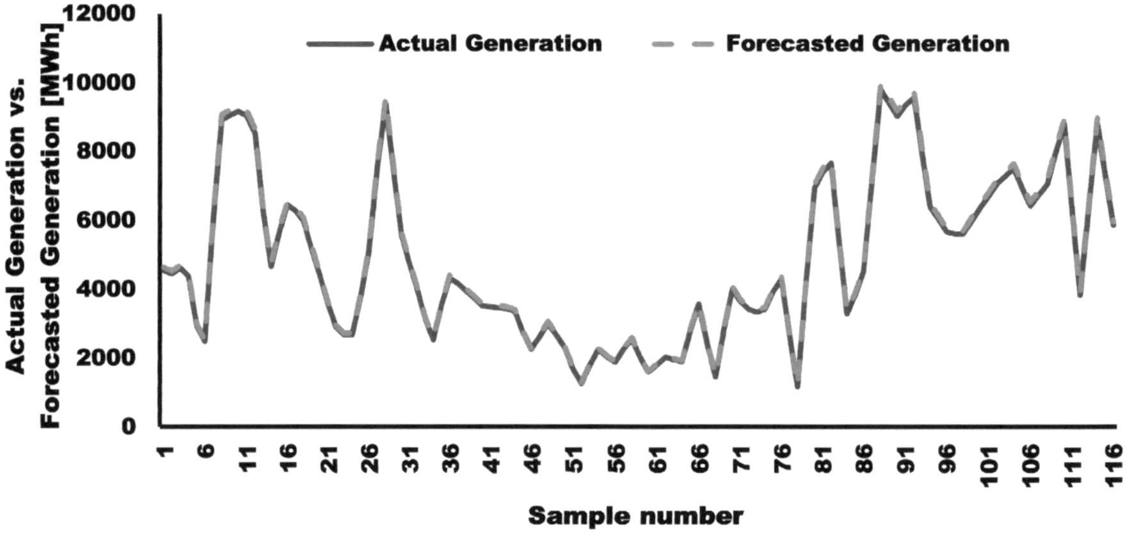

Figure 8. Actual generation vs. forecasted generation with augmented data.

The forecast in Figure 8 covers exactly the same period between 03.09.2007-18.02.2012, as in Figure 5.

The Augmented Dickey Fuller Test revealed again that the whole augmented time series is stationary as well as the first 80 % of the data that has been used for training the multiple input CNN. The rest of 20 %, that has been used for model validation, is not stationary.

In this case, the MAE=87.7422 MWh, the MBE=87.7422 MWh, the MAPE=1.8164 %, and the RMSE=96.8442 MWh.

In order to have a better idea of the comparison between the case with non-augmented and the case with augmented data, refer to Table 1.

Table 1. Comparison between the prediction errors obtained with non-augmented and with augmented data sets.

Analysis	MAE [MWh]	MBE [MWh]	MAPE [%]	RMSE [MWh]
CNN with non-augmented data	89.7001	89.7001	1.8849	100.6000
CNN with augmented data	87.7422	87.7422	1.8164	96.8442
Improvement [%]	2.2	2.2	3.7	3.8

4. Conclusions

In this work, a multiple input (multi-channel) CNN for wind power generation forecasting, together with a GAN for data augmentation, have been introduced. The training of the CNN has been done simultaneously with historical data related to wind speeds and power generation. A comparison between the errors obtained with non-augmented and with augmented training data has been conducted, as well. The model proved its robustness and reliability when renouncing at data normalization, prior ingestion, and when renouncing to batch normalization. As can be seen in Table 1, the accuracy of the multi-channel CNN was 98.1151 % when trained with non-augmented data, and 98.18 % when trained with augmented data. The general improvement when using augmented training data is of 2.975 % which impacts high when applied to the MWh domain. Future research should be aimed at improving this neural model in order to stimulate its adoption in the industry. That is related to various aspects such as cloud deployment or the quantification of the relationship between the available training sample number and forecast precision.

At the same time, the GAN data augmentation proves its potential also when the initial data set is not good enough (does not contain enough observations) to obtain a forecast accuracy of at least 90 %. This could have many reasons, among which GDPR regulations play a key role. As such, the GAN data augmentation represents a very good solution also when coping with official restrictions imposed by the EU or similar regulators or when the training data contains outliers that have to be removed and, then, replaced by correct values.

References

[1] https://www.databricks.com/solutions/accelerators/demand-forecasting [Accessed 08.04.2024]
[2] https://www.capstoneinfrastructure.com/our-businesses/operating-facilities/ErieShoresWindFarm [Accessed 10.04.2024]
[3] N. Pattanajak and H. Malekmohamadi, "Improving a 3-D Convolutional Neural Network Model Reinvented from VGG16 with Batch Normalization," 2019 IEEE SmartWorld, Ubiquitous Intelligence & Computing, Advanced & Trusted Computing, Scalable Computing & Communications, Cloud & Big Data Computing, Internet of People and Smart City Innovation (SmartWorld/SCALCOM/UIC/ATC/CBDCom/SCI), pp. 45-50, Leicester, UK, 2019.
[4] https://www.mathworks.com/help/matlab/ref/conv2.html [Accessed 17.04.2024]

IOP Conf. Series: Materials Science and Engineering **1320** (2024) 012003 doi:10.1088/1757-899X/1320/1/012003

The influence of the dielectric material on the planar antennas' parameters

Sabin Dușa[1*], Claudia Constantinescu[1], Adina Giurgiuman[1], Sergiu Andreica[1], Razvan Gliga[1], Claudia Pacurar[1*]

[1] Department of Electrotechnics and Measurements, Technical University of Cluj-Napoca, Cluj, Romania
*E-mail: Sabin.Dusa@campus.utcluj.ro; Claudia.Pacurar@ethm.utcluj.ro

Abstract. Patch antennas made in planar electromagnetic technology have found applicability in the last few years in a varied range of fields, but predominantly in wireless communications and telecommunications. This type of antenna is popular due to their reduced dimensions, their ease of manufacture and adaptability to integration into various devices. In this paper the design of a compact antenna operating under 6 GHz, for use in 5G applications, is proposed. For this purpose, a planar monopole antenna with two arms, fed by a coplanar microstrip line, printed on different types of dielectric substrates was numerically modelled and constructed and the influence of the dielectric material on the antenna parameters was analysed in order to further design some array and MIMO (multiple input - multiple output) antennas. The parameters considered to be studied to reach some optimum structures operating at high performances were the S parameters, directivity and gain. For this purpose, four sets of planar monopole antennas were designed and modelled using a commercial software used for high frequency analysis. The results of the numerical analysis were validated also through experimental measurements.

1. Introduction

The rapid evolution of technology in wireless communications and telecommunications field generated a major interest for miniature antennas made in planar electromagnetic technology. This evolution is motivated by the need to operate at higher and higher frequencies, in the order of GHz, while also aiming to reduce the size of these components down to the micron scale. These trends involve significant changes to both the antennas and the entire device. Consequently, circuits operating at high and very high frequencies play a crucial role in optimizing the transmission/reception power efficiency, as a larger bandwidth is sought for the same resonant frequency in order to obtain devices operating with superior performances.

With the increased demand for wireless services and the emerging 5G communications accompanied by the Internet of Things (IoT) concept, a huge amount of data is expected to be exchanged between users of these technologies. One of the main elements of the systems that are used to serve the requirements of 5G technology is the antenna. The antenna can be designed as a single element or constructed from multiple elements. Most antenna models from the specialized literature are unique elements [1]-[5]. However, single-element antennas have limited capabilities from a gain and bandwidth perspective. The research presented in this article seeks to overcome these limitations by using dielectric materials hoping to increase the gain and reach a wider bandwidth.

In Romania and in Europe, the frequency bands of 800 MHz, 900 MHz, 1800 MHz, 2100 MHz, 2600 MHz and 3500 MHz are used for mobile communications. Electromagnetic waves with frequencies in bands considered "low", below 1 GHz (700 MHz, 800 MHz, 900 MHz) are favourable for covering large areas, over long distances, they propagate well inside buildings, but the possible data transmission speed is limited [6]-[8]. On the other hand, waves with frequencies in "medium" bands (2600 MHz, 3400-3800 MHz) propagate over shorter distances and do not penetrate well inside buildings but allow high data transmission speed [9]-[11].

The industry of high-frequency micro-antennas made in planar electromagnetic technology has developed exponentially with the extraordinary advances in communications in the last few years, where the need to provide certain types of antennas, which are characterized by small dimensions, high efficiency and high frequency transmissions has become a necessity.

In general, a microstrip antenna is known as a "printed antenna" and its patch usual shapes are square, rectangular, circular, and oval, but other shapes are also used, each with their own unique characteristics and functionality.

2. The planar monopole antenna with two arms

2.1. Design and high frequency modelling

In this study, a monopole microstrip antenna design with two arms, composed of: copper conductive patch - as the main body of the antenna, insulating dielectric substrate, respectively ground plane is introduced. The analysed antenna configuration is presented in Figure 1.

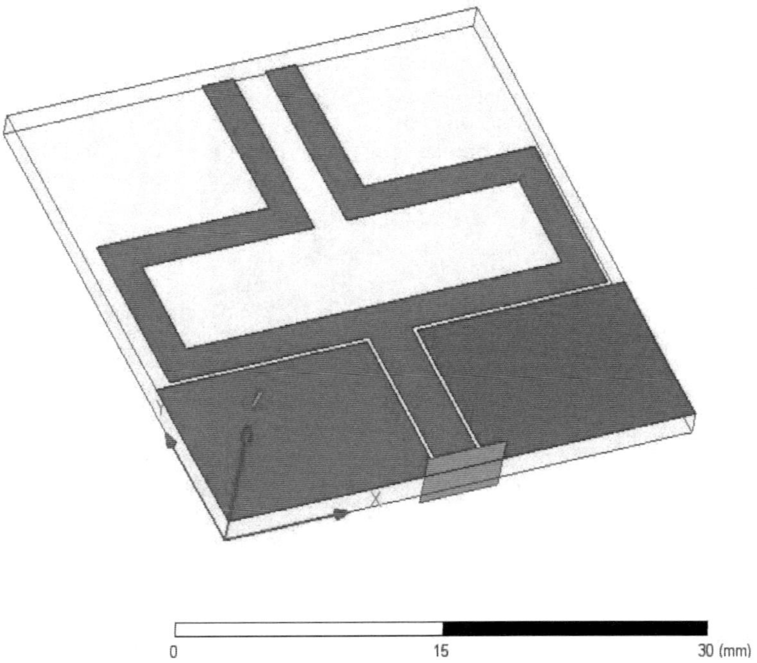

Figure 1. Antenna structure.

The antenna is fed with a coplanar waveguide and the conductor is placed on a Rogers 3010 substrate with the thickness of 1.28 mm and a dielectric permittivity of ε_r= 10.2. The geometrical dimensions of the substrate are 30x30x1.28 mm. On the left and right of the microstrip line feeding the antenna there are copper layers with dimensions 13.4x9.75 mm as grounds for the structure (GND1, GND2) for the main body of the antenna, they can be seen in Figure 2.

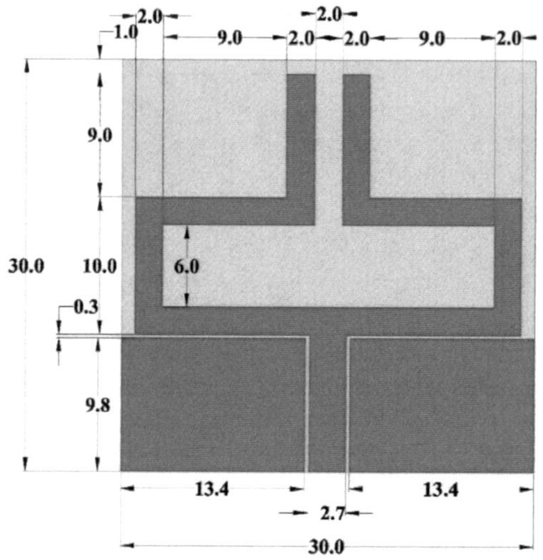

Figure 2. Dimensions of the designed antenna.

The 3D numerical modelling and functional simulations of the antennas were made with the help of the software program ANSYS-High Frequency Structural Simulator (HFSS), in the frequency range 1GHz–6 GHz.

The targeted objectives were the high-frequency analysis of the specific antenna parameters: resonance frequency; bandwidth; gain and directivity.

The resonance frequency is determined from the S parameters representation as can be seen in Figure 3.

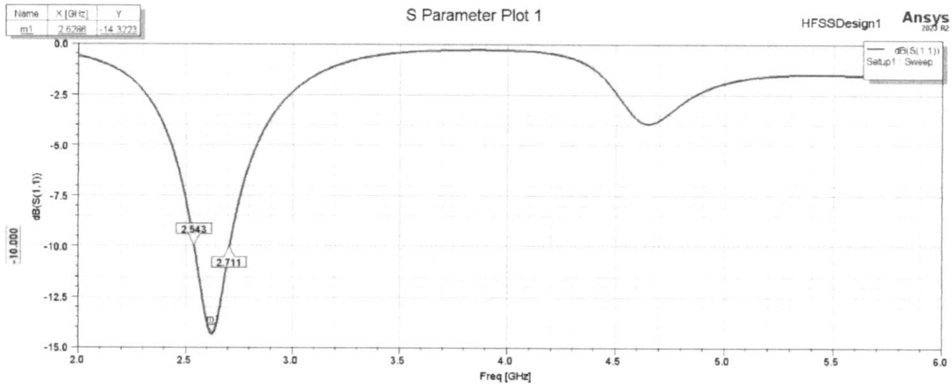

Figure 3. Graphical representation of S11

According to the graph, the resonance frequency is determined to be at 2.628 GHz at an amplitude of the S11 parameter of -14.32 dB. The bandwidth of the antenna can be calculated as the difference between the two points of the S parameters intersection the -10 dB line, thus observing that the first point in this case is f_1= 2.54 GHz and f_2=2.71 GHz, it can be stated that its value in this case is 14dB. After determining the central frequency f_c with formula (1), the fractional bandwidth can be also can calculated with formula (2) to be 6.48%.

$$f_c = \frac{f_1+f_2}{2} \tag{1}$$

$$\mathrm{FBW}=\frac{f_2-f_1}{f_c}\cdot 100 \tag{2}$$

Other parameters of interest that were determined, are: peak directivity=3.88, peak gain=2.56, radiation efficiency=0.65 and front to back ratio=1.025.

An antenna's gain is a measure of its ability to focus or direct signal energy in a particular direction. It is usually expressed in decibels (dB) and is the ratio of the power emitted in a given direction by the antenna of interest to the power a reference antenna, such as an ideal antenna, would emit. It can be seen that the antenna emits more in the vertical direction, above the patch and its maximum gain is 2.56 as can be seen in Figure 4.

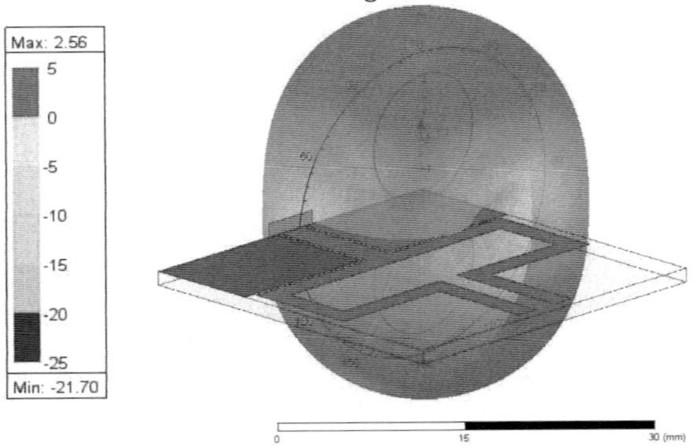

Figure 4. Graphical representation of the gain.

Directivity represents the ability of an antenna to direct as much power as possible in a certain direction and reduce the power radiated in other directions. The higher the directivity, the more effective the antenna is in directing the signal to or from a specific point or area in space. Directivity is closely related to antenna gain and is one of the key characteristics of antenna performance. In Figure 5 is represented the polar 3D diagram of the antenna's directivity.

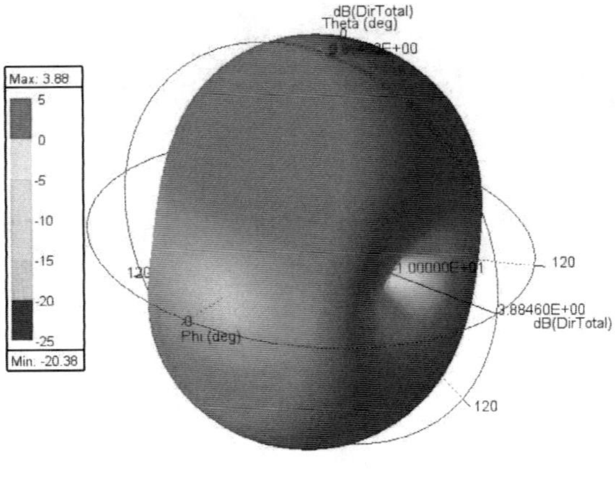

Figure 5. Graphical representation of the antenna's directivity

2.2. Construction and experimental measurement of the antenna

The antenna was built using the LPKF ProtoMat S103 plotter by importing the designed geometry in the ANSYS High Frequency Structure Simulator (HFSS) simulation module into the dedicated plotter design software, LPKF Circuit Pro (Figure 6).

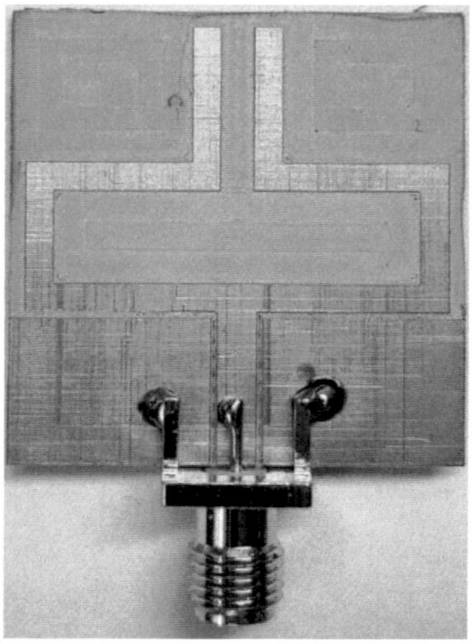

Figure 6. Practically constructed antenna

After the construction of the structure on the Roger 3010 substrate, the S-parameters were measured using the ZLN 20 Vector Network Analyzer working in the frequency range 5 kHz-20 GHz, and the experimental stand is shown in Figure 7. The S parameters that were experimental measured can be seen on the VNA display.

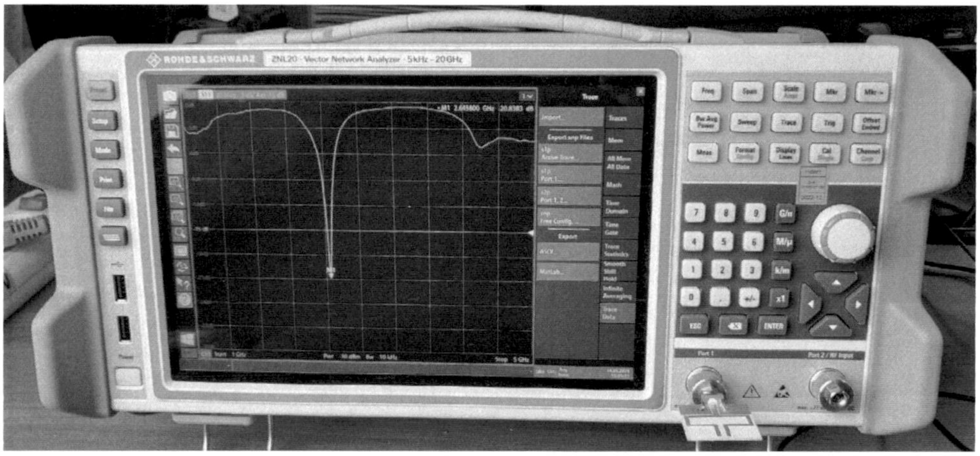

Figure 7. Experimental stand

In Figure 8, the S parameters of the designed antenna experimentally measured for the constructed antenna are presented in order to be compared with the numerical modelling results for the S parameters from Figure 3. After comparing the results obtained by numerical modelling and experimentally measuring the S parameters, it is found that the resonant frequency obtained by simulating with the help of the HFSS program is 2.628 GHz, and the experimentally measured one is 2.645 GHz. This indicates that the model simulated in the HFSS program is in accordance with the real behaviour of the antenna. Such agreement between simulation and measurement confirms the validation of the designed model.

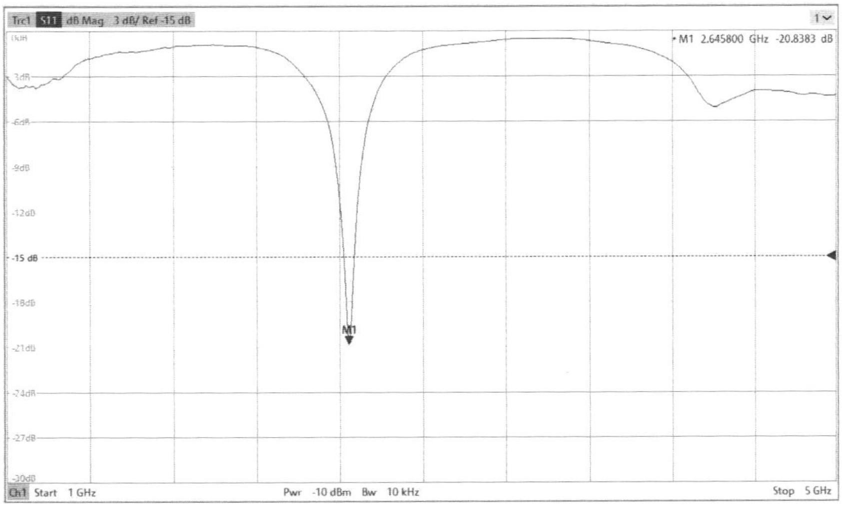

Figure 8. The S parameters from the experimental testing of the antenna's structure

3. The analysis of the dielectric material influences on the planar antennas' parameters

Dielectric material selection in antenna design significantly affects antenna characteristics, including performance, radiation pattern, impedance distribution, and bandwidth. The temperature and frequency stable dielectric constant of dielectric materials makes them ideal for bandpass filters, microstrip patch antennas and voltage-controlled oscillators [3], [6], [8].

In order to determine how different dielectric materials and their geometric dimensions affect the operation of the antenna, four sets of antennas were designed in which different dielectric materials were considered for the substrate: Roger3010, Roger5880; Roger4350; Roger TMM10. The material properties for the analysed dielectrics can be found in Table 1.

Table 1. Dielectric materials properties.

Material type	Relative permittivity	Thickness[mm]
Roger 3010	10.2	0.13 0.25 1.28
Roger TMM 10	9.8	0.381 0.635 0.762
Roger 4350	3.66	0.25 0.51 0.76
Roger 5880	2.2	0.127 0.508

3.1. Analysis of resonance frequency and bandwidth variation using Roger 3010 material as substrate

Starting from the initial structure, described above constructed on Roger 3010 dielectric substrate, with the relative permittivity 10.2, we designed a few structures considering different dielectric thickness as can be seen in Table 2.

Table 2. Dielectric thicknesses.

Material type	Relative permittivity	Thickness[mm]
Roger 3010	10.2	0.13 0.25 1.28

Considering the dielectric thicknesses of 0.13 mm and 0.25 mm and the geometry of the antenna and the grounds the same, the S parameters representations from Figure 9 were determined.

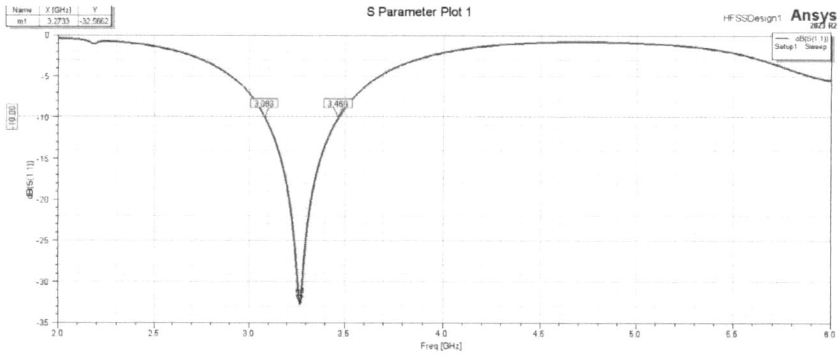

a) 0.25 mm

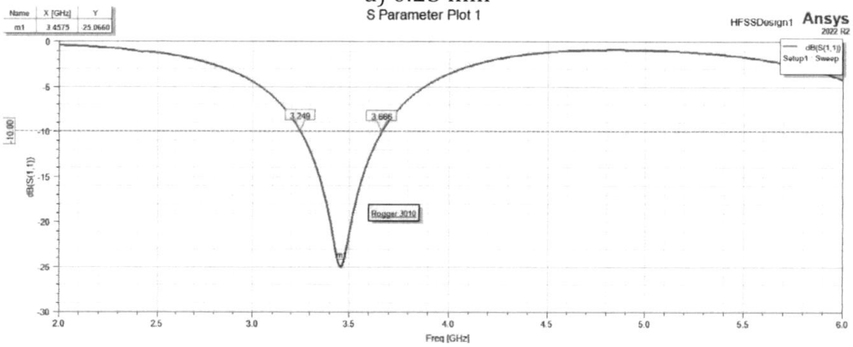

b) 0.13 mm

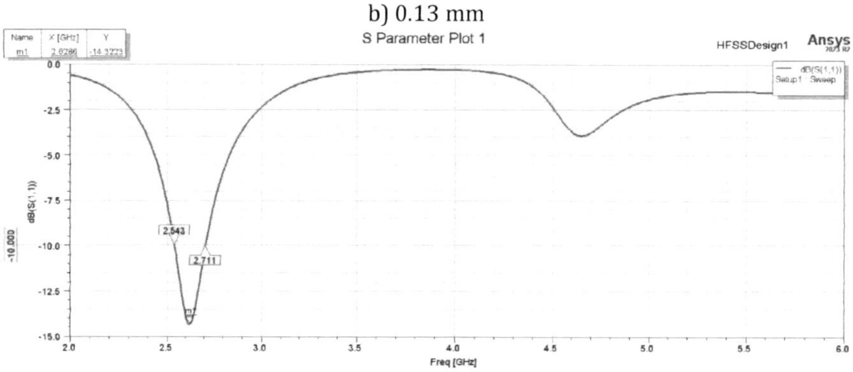

c) 1.28 mm

Figure 9. S parameters representation for the antennas having as dielectric Rogers 3010

According to the graph for the structure with the dielectric thickness of 0.25 mm, the resonance frequency is found at 3,273 GHz, where S11 has an amplitude of -32,56 dB. In this case, after determining f_1= 3.08 GHz and f_2=3.46 GHz, it can be stated that the central frequency f_c has a value of 3.27 GHz calculated with formula (1). The fractional bandwidth can be also can calculated with formula (2) to be 11.62 %.

In the case of the structure with the dielectric thickness of 0.13 mm, the resonant frequency is close to the one obtained in the previous case, namely at 3.45 GHz. The central frequency in this case is calculated to be 3.45 GHz, while the fractional bandwidth is 12.17 %. All the results can be seen in Table 3 for a better comparison.

Table 3. Resonance frequency and bandwidth variation with the substrate thickness for Rogers 3010.

Material type	Relative permittivity	Thickness[mm]	Resonance frequency [GHz]	S11 amplitude [dB]	Fractional bandwidth [%]
Roger 3010	10.2	0.13	3.45	-25.06	12.17
		0.25	3.27	-32.56	11.62
		1.28	2.62	-14.30	6.48

In Figure 10 the gain representation for the gain of the two considered structures having as dielectric Rogers 3010 can be observed.

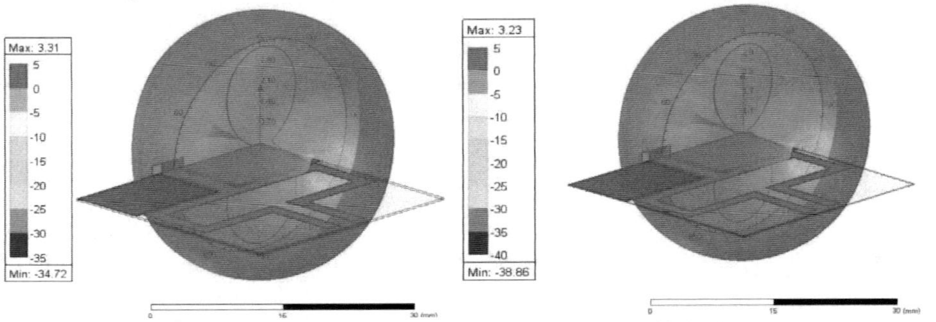

a) 0.25 mm b) 0.13 mm

Figure 10. Gain representation for the antennas having as dielectric Rogers 3010

The peak directivity for the antenna with dielectric thickness of 0.25 mm is 2.166, the radiation efficiency is 0.99031, while the front to back ratio is 1.0053. In the case where the thickness is reduces at 0.13 mm, the peak directivity is smaller at a value of 2.1147, the radiation efficiency is 0.99458, while the front to back ratio is 1.005. After analysing these results, it can be stated that the resonance frequency value and the bandwidth are inversely proportional influenced by the substrate thickness (Table 3), thus a reduction of the substrate thickness leads to an increase of the bandwidth. Also, the resonance frequency reaches a higher value, thus in designing an antenna to operate at a certain frequency value, modifying the dielectric thickness is an important tool.

3.2. Analysis of resonance frequency and bandwidth variation using Rogers TMM10 material as substrate

For this study, the dielectric thickness of RogersTMM10 is modified. This dielectric has a lower relative permittivity (ε_r= 9.8) than Rogers 3010. The dielectric thicknesses considered for this study can be seen in Table 4.

Table 4. Dielectric thicknesses for Rogers TMM10.

Material type	Relative permittivity	Thickness[mm]
Roger TMM 10	9.8	0.381 0.635 0.762

The S parameters for the three considered cases are presented in Figure 11.

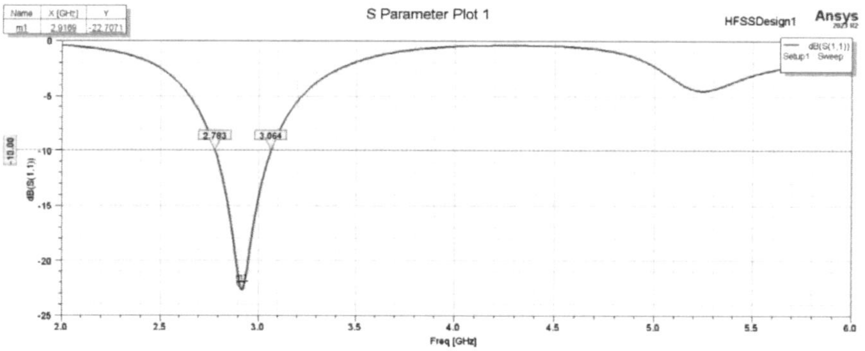

a) 0.762 mm

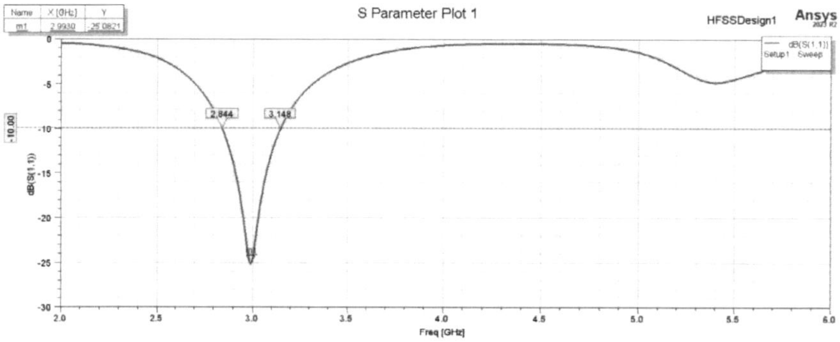

b) 0.635 mm

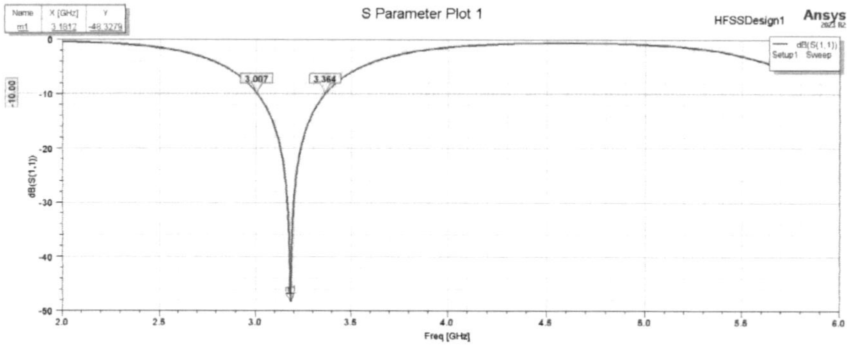

c)0.381 mm

Figure 11. S parameters representation for the antennas having as dielectric Rogers TMM10

IOP Conf. Series: Materials Science and Engineering **1320** (2024) 012003 doi:10.1088/1757-899X/1320/1/012003

The resonance frequency and fractional bandwidth for these cases are presented in Table 5.

Table 5. Resonance frequency and bandwidth variation with the substrate thickness for Rogers TMM10.

Material type	Thickness[mm]	Resonance frequency [GHz]	S11 amplitude [dB]	Fractional bandwidth [%]
Rogers TMM 10	0.381	3.81	-48.32	11.32
	0.635	2.99	-25.08	10.03
	0.762	2.91	-22.7	9.62

It can be stated that the conclusion drawn for the Rogers 3010 substrate remains the same, namely the decrease of the dielectric increases the fractional bandwidth and makes the resonance frequency have higher values. Also, the decrease of the relative permittivity makes the bandwidths increase at thicker dielectrics.

Figure 12 gives as the representations of the gain of the antenna. After determining the maximum values for the gain, it can be stated that it is not significantly influenced by the change of the dielectric thickness for the considered cases, being a little bit smaller for the thinner dielectric, but maintaining the same allure. Comparing the results with the ones for the Rogers 3010 dielectric, it can be said that the gain increases for the TMM10 structures. The directivity for the antenna with dielectric thickness of 0.762 mm is 2.32, radiation efficiency is 0.981212, and front to back ratio is 1.012. The directivity for the antenna with dielectric thickness of 0.635 mm is 2.2918, radiation efficiency is 0.98435, and front to back ratio is 1.0101. The directivity for the antenna with dielectric thickness of 0.381 mm is 2.2021, radiation efficiency is 0.99344, and front to back ratio is 1.0076.

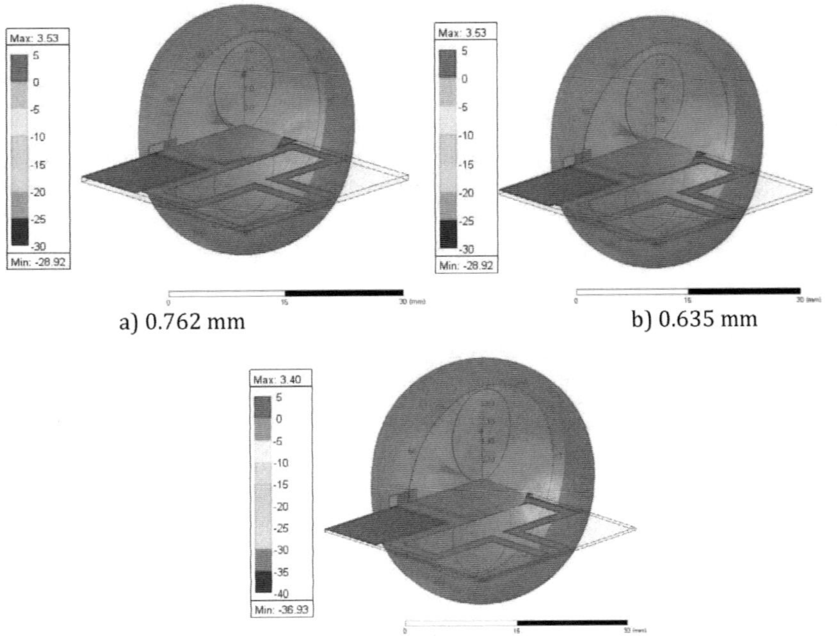

a) 0.762 mm b) 0.635 mm

c)0.381 mm

Figure 12. Gain representation for the antennas having as dielectric Rogers TMM10

IOP Conf. Series: Materials Science and Engineering **1320** (2024) 012003 doi:10.1088/1757-899X/1320/1/012003

3.3. Analysis of resonance frequency and bandwidth variation using Rogers 4350 material as substrate

The authors also considered some dielectrics with smaller relative permittivity namely Rogers 4350 which has ε_r=3.55 and after consulting the datasheets for this type of dielectric they considered the standard dimensions 0.76 mm, 0.51 mm, 0.25 mm for the analysis. Thus, in Figure 13, the S parameters for these structures are presented.

The conclusions remain the same as for the two previous cases. Analysing the results, and considering the resonance frequency, it can be stated that the thinner dielectrics and dielectric materials with smaller relative permittivity can improve the antenna performances, as it can be seen in Table 6.

Table 6. Resonance frequency and bandwidth variation with the substrate thickness for Rogers 4350.

Material type	Thickness[mm]	Resonance frequency [GHz]	S11 amplitude [dB]	Fractional bandwidth [%]
Rogers 4350	0.25	3.64	-29.45	14.01
	0.51	3.62	-13.71	13.9
	0.76	3.38	-29.4	12.42

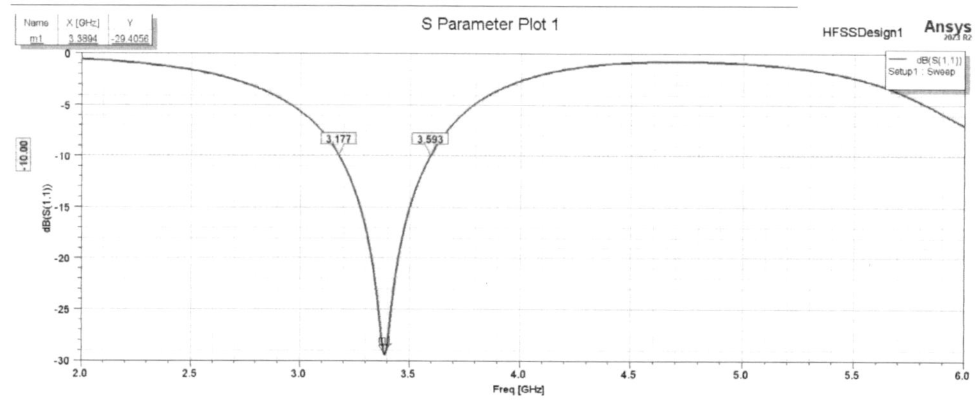

a) 0.76 mm

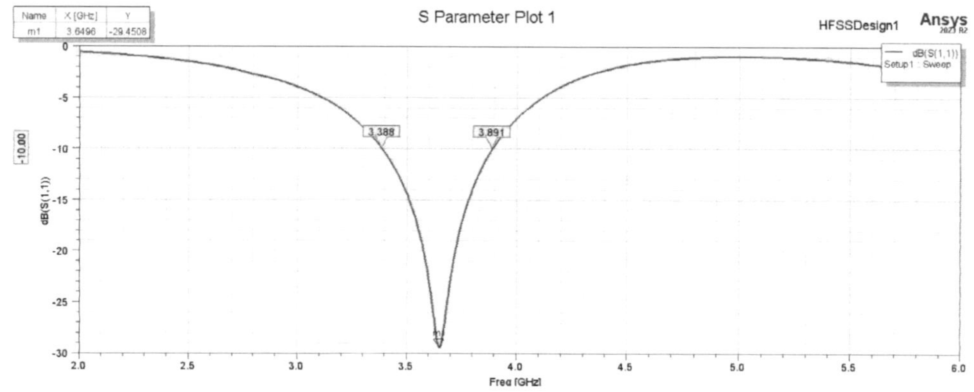

b) 0.51 mm

IOP Conf. Series: Materials Science and Engineering **1320** (2024) 012003 doi:10.1088/1757-899X/1320/1/012003

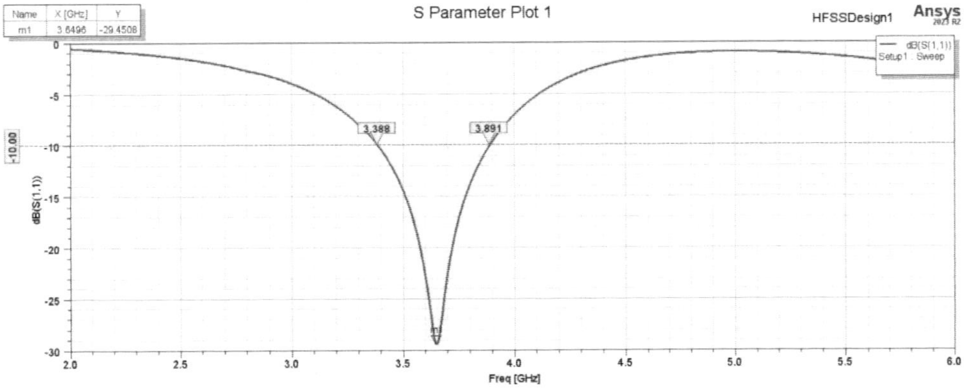

c)0.25 mm

Figure 13. S parameter representation for the antennas having as dielectric Rogers 4350

In Figure 14 the gain for the considered structures is represented. The gain decreases from the previous cases, while the efficiency and front to back ratio are not significantly influenced. The peak directivity is 2.1403 for 0.76 mm dielectric thickness, 2.1037 for 0.51 mm thickness, and 2.087 for the 0.25 mm.

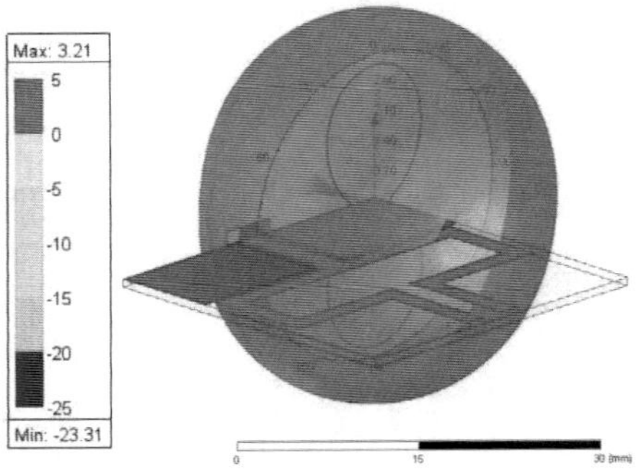

a) 0.76 mm

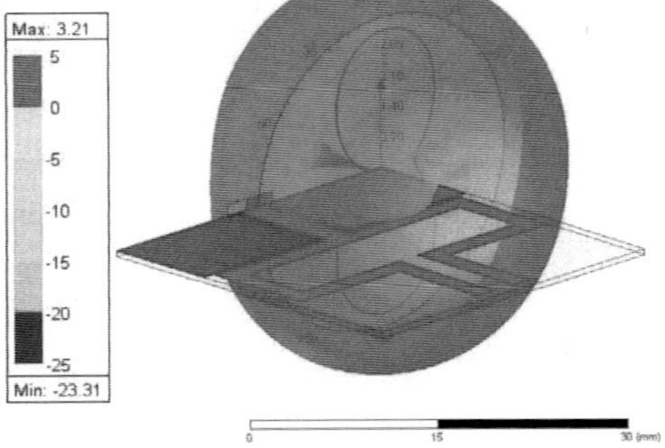

b) 0.51 mm

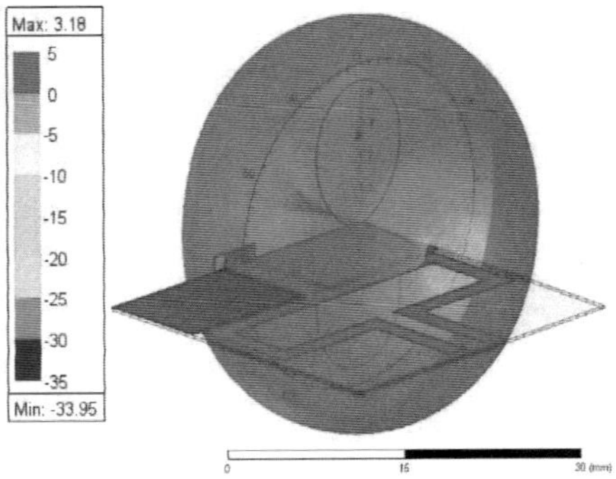

c)0.25 mm

Figure 14. S parameter representation for the antennas having as dielectric Rogers 4350

IOP Conf. Series: Materials Science and Engineering **1320** (2024) 012003 doi:10.1088/1757-899X/1320/1/012003

3.4. Analysis of resonance frequency and bandwidth variation using Rogers 5880 material as substrate

The studied case is the one for the Rogers 5880 dielectric with a relative permittivity of 2.2, considering two thicknesses namely 0.127 mm, 0.508 mm.

The S parameters are represented in Figure 15 while in Figure 16 the gain can be observed. The resonant frequency is at lower values while the fractional bandwidth is smaller than the one obtained for the other dielectrics (Table 7). The peak directivity is 2.2718 for the first structure and 2.4068 for the second one. The front to back ratio increased in comparison to the other analysed structures, reaching 2.0306 and 1.9463 respectively, fact which is not good.

Table 7. Resonance frequency and bandwidth variation with the substrate thickness for Rogers 5880.

Material type	Thickness[mm]	Resonance frequency [GHz]	S11 amplitude [dB]	Fractional bandwidth [%]
Rogers 5880	0.127	2.28	-22.08	3,5
	0.508	2.17	-19.75	3.22

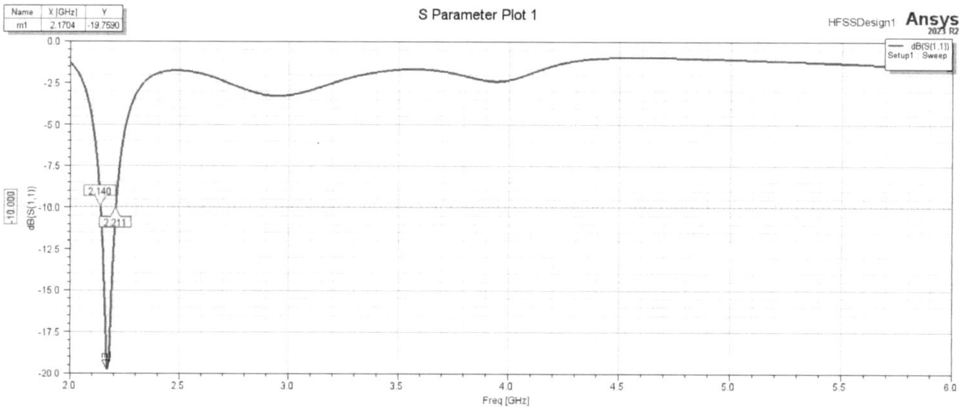

a) 0.508 mm

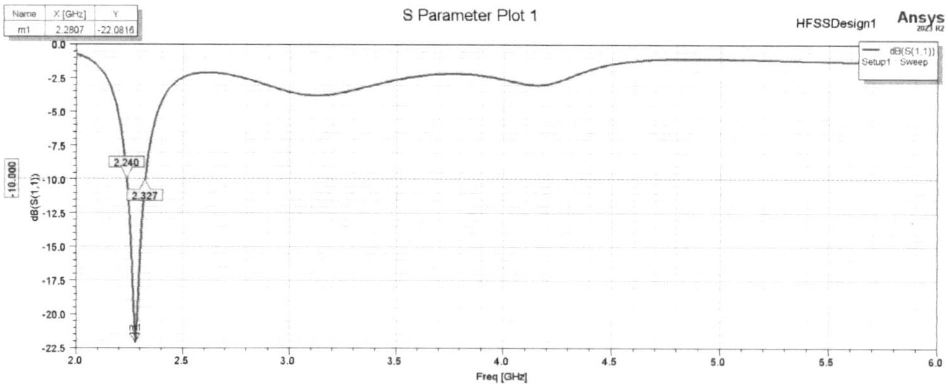

b) 0.127 mm

Figure 15. S parameter representation for the antennas having as dielectric Rogers 5880

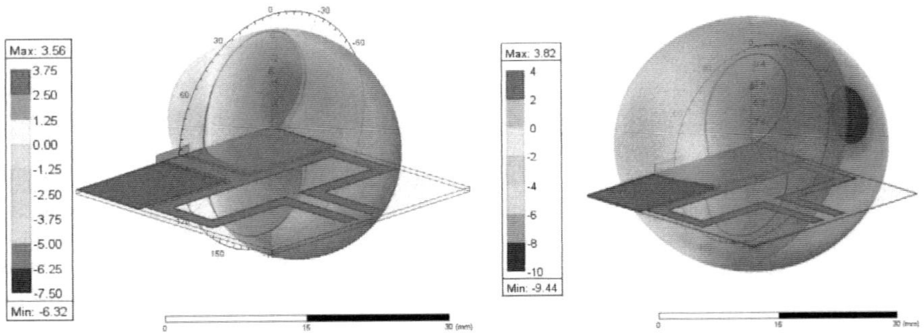

a) 0.508 mm b) 0.127 mm

Figure 16. Gain representation for the antennas having as dielectric Rogers 5880

4.Conclusions

The geometrical dimensions and the dielectric materials used for the construction of an antenna greatly influence its operational characteristics such as bandwidth, resonant frequency, gain and directivity. It has been observed by considering different dielectric materials and by modifying their thickness that no matter the dielectric chosen, once the thickness is smaller the fractional bandwidth increases. The type of dielectric chosen influences the resonant frequency, thus when in need of an antenna to work at a certain frequency an important parameter to vary is the dielectric material. Because we considered the pre-existent dimensions of the dielectrics, we cannot compare one on one these results, thus based on the producer offer the designers must optimize their product. The values of the gain obtained are between 3.21 and 3.83, but the highest value was obtained for the Rogers 5880 material, followed by Rogers TMM10, while the highest fractional bandwidth was obtained for Rogers 4350, thus the designer must consider the applications of the antenna and reach a compromise regarding these characteristic parameters.
All the analysis, performed for this article, are needed for the author's further research work aiming at the design of array antennas, respectively MIMO antennas dedicated to 5G and 6G applications that will be constructed on properly dielectric material.

References

[1] Chen, Q.; Lin, H.;Wang, J.; Ge, L.; Li, Y.; Pei, T., Single ring slot-based antennas for metal-rimmed 4G/5G smartphones. IEEE Trans. Antennas Propag. 2018, 67, 1476–1487.

[2] Foschini, G.J.; Gans, M.J. On limits of wireless communications in a fading environment when using multiple antennas. Wirel.Pers. Commun. 1998, 6, 311–335.

[3] Racasan, Adina; Munteanu, C.; Topa, V.; Pacurar, Claudia; Hebedean, Claudia, Electromagnetic Interferences Suppresion in Planar Integrated Devices, Proc. of the 2014 International Symposium on Electromagnetic Compatibility, EMC Europe 2014, pp. 940-945, 2014.

[4] Levitas, B.; Drozdov, M.; Naidionova, I.; Jefremov, S.; Malyshev, S.; Chizh, A., UWB system for time-domain near-field antenna measurement. In Proceedings of the 2013 European Microwave Conference, Nuremberg, Germany, 6–10 October 2013; pp. 388–391.

[5] Cretu, Mihaela; Ciupa, R. V.; Cretu, T., Assessment of the Electric Field Generated by Multilayered Coils during MS, ATEE 2013, 23-25 May, Bucharest, Romania, pp. 207-210, 2013, WOS:000332928500090

[6] Constantinescu, Claudia; Pacurar, Claudia; Giurgiuman, Adina; Munteanu, C.; Andreica, S.; Gliga, M., High Gain Improved Planar Yagi Uda Antenna for 2.4 GHz Applications and Its Influence on Human Tissues. Applied Sciences-Basel, vol. 13, no. 11, 6678, ISSN: 2076-3417, DOI10.3390/app13116678, WOS:001005579400001, IF: 2.7, May 2023.

[7] Hussain, R.; Alreshaid, A.T.; Podilchak, S.K.; Sharawi, M.S. Compact 4G MIMO antenna integrated with a 5G array for current and future mobile handsets. IET Microw. Antennas Propag. 2017, 11, 271–279.

[8] Racasan, Adina; Munteanu, C.; Topa, V.; Pacurar, Claudia; Hebedean, Claudia; Minimization of the Equivalent Parallel Capacitance in Planar Magnetic EMI Filters, Proceedings of the International Conference and Exposition on Electrical and Power Engineering, 7th edition, EPE 2012, pp. 519-524, 2012.

[9] Hassan, M.M.; Rasool, M.; Asghar, M.U.; Zahid, Z.; Khan, A.A.; Rashid, I. A novel UWB MIMO antenna array with band notch characteristics using parasitic decoupler. J. Electromagn. Waves Appl. 2019, 34, 1225–1238.
[10] Gliga, M.; Racasan, Adina; Munteanu, C.; Andreica, S.; Pacurar, Claudia; Topa, V.; Constantinescu, Claudia, The Influence of Ferrite on the Spiral Inductors Inductance used for the Design of Wireless Power Systems, Proceedings - 2017 International Conference on Modern Power Systems, MPS 2017, 2017.
[11] ANCOM – Totul despre tehnologia 5G (All about 5G Technology) – Frecventa pentru comunicatiile mobile, inclusive 5G (The frequency of mobile communication, including 5G)- https://infocentru.ancom.ro/tehnologia-5g-2/#1610461299432-9fb7ff96-dc02

IOP Conf. Series: Materials Science and Engineering **1320** (2024) 012004 doi:10.1088/1757-899X/1320/1/012004

Impact of Structural Design of Knitted Fabrics on Shielding from Radiation of Wi-Fi Technologies

M Grecka[1*], I Baltina[1], J Blums[2] and J Eidaks[3]

[1] Institute of Architecture and Design, Riga Technical University, Riga, Latvia
[2] Technical Physics Institute, Faculty of Natural Sciences and Technology, Riga Technical University, Riga, Latvia
[2] Institute of Photonics, Electronics and Electronic Communications, Faculty of Computer Science, Information Technology and Energy, Riga Technical University, Riga, Latvia

*E-mail: Marianna.Grecka@rtu.lv

Abstract. The goal of the study is to determine and compare the attenuation of electromagnetic waves at 2.4 - 2.5 GHz frequency between plain single jersey and 1x1 rib knitted fabric. For knitting the samples, three different types of metal-contained yarns, their combinations, and cotton yarn were used. Two horn antennas and a network analyzer were used to characterize the transmission of the samples in the frequency range of 2.4–2.5 GHz. Obtained results show, that in mentioned frequency range single jersey knitted samples attenuate better than the rib samples.

1. Introduction

Continuous use of communication technologies (e.g. smartphones or wireless internet) pollutes the environment with electromagnetic (EM) fields (non-ionizing radiation) that affect human health negatively. Even medical scientists created such a disease designation as electromagnetic sensitivity. Individuals dealing with this issue describe such unpleasant feelings symptoms as sleep disturbances, headache, nervousness, tiredness, inability to concentrate, heart disease, redness of the skin, eye disease, depression, back pain, circulatory problems, and tremors [1, 2].

The causes of human health disorders have been researched in depth – at the cellular level. Since the 1970s, many experiments have been conducted to prove the adverse effect of EM fields on live creatures, such as animals and human cells. In publications for the last decade, specialists and scientists have summarized enough research to review on topics, such as the effects of radiofrequency EM field exposure on male fertility, pregnancy, and birth outcomes [3], oxidative stress [4], cellular DNA damage, changes in non-steroid hormone levels, heat shock protein induction, and neurological and neuropsychiatric effects [5].

Basing on similar review on such topics, International Commission on Non-Ionizing Radiation Protection had presented reasoned guidelines for limiting exposure to electromagnetic fields (EMFs) that will provide a high level of protection for all people against substantiated adverse health effects from exposures to both short- and long-term, continuous and discontinuous radiofrequency EMFs [6].

A number of precautionary measures can mitigate the harmful effects of electromagnetic fields on the human body. Beside shortening the exposure time to the source of radiation, e.g., by using the equipment or device as needed and increasing the distance from the equipment or device during usage, people can protect themselves by using special EM shields and screens [7]. The shields can be made from a metal mesh of planes and textile materials, e.g., woven or knitted fabrics and non-woven materials, with metal-contained coatings for shielding. As well, these textile materials may consist of various structures' yarns or threads – from different traditional fibers blended or twisted with metal (copper, stainless steel, silver, gold), and/or metal-plated fibers [8-13].

Many researchers turn to the study of shielding textiles, as well as knitted ones. For example, in study [14] there were determined the shielding properties of single jersey samples made of acrylic and steel fibre yarns on electronic flatbed knitting machine, gauge 7. Six samples of different density were measured in the wavelength from 750 to 3000 MHz in vertical and horizontal plane of polarization. The samples shielded better in the vertical plane of polarization, but sample with highest density (6 course/1 cm and 4.5 wales/1 cm) showed the best results in both planes of polarization.

In another study [15], the shielding effectiveness (SE) of charmeuse knitted structure samples were described. Knitted fabric of polyamide yarn with Ag coating were tested at 0.9, 1.8, 2.1 and 2.4 GHz frequency before and after ten cycles of wet and dry cleaning. Within the increasing of frequency, the SE increased. But by increasing the number of cleaning cycles, the shielding properties of the samples reduced.

In the study by Palanisamy et. al. [16], the single jersey and double jersey (1 × 1 rib) fabrics of silver-coated yarn, made with two different patterns and three levels of stitch densities, were measured for EM shielding effectiveness in the frequency 30 MHz to 1.5 GHz. The double jersey sample set showed higher shielding ability compared to single jersey fabrics at 1.5 GHz.

Evaluating the mentioned research in some way similar methods, frequencies, and parameters of samples are used, but there is no studied knitted fabrics with combined types of yarns, like in our research. The goal of the study is to clarify and compare the attenuation of electromagnetic waves in the frequency range of 2.4-2.5 GHz of single jersey and 1x1 rib pattern made knitted textiles consisting of metal-contained and cotton yarns. The mentioned spectrum is the most widely used of the bands available for Wi-Fi [17]. There are a lot of everyday devices that use Wi-Fi technologies, such as computers, laptops, smartphones, tablets, smart TVs, printers, digital audio players, digital cameras, cars and drones, etc., respectively, exposure to such devices magnetic field irradiation is inevitable and can hardly be avoided. Taking into account the negative aspect of the exposure it is important to protect the human, which can be achieved by using electromagnetic shielding textile, such as clothing of metal-contained knitted textiles.

2. Materials and methods

2.1. Materials

For preparation of knitted samples, there were three types of metal-contained yarns and cotton yarn. The composition and parameters of selected yarns are shown in the Table 1. There were different textures of metal-contained yarns selected, taking into account such factors as the closest conformity to traditional textile yarns, the structure and chemical consistency of the yarns, and the possibility of using the yarns in a knitting machine. The steel staple fiber of yarn A is the finest (0.01 mm) among the other metal filaments (0.03 mm) of the yarns. The resistance of 10 cm long yarn, respectively, its metal part, was measured. The highest resistance is observed to yarn A, which is explained by its not-uniform structure of metal components. The steel staple fibers are short and connect to each other only by twisting, which increases the resistance. While the resistance of yarns B and C is much smaller than that of yarn A, because these yarns have continuous metal filaments in their composition.

Table 1. Metal contained yarns for knitted samples preparation

Code of the yarn	Yarn A	Yarn B		Yarn C	
Type	Single spun yarn	2 plied yarn formed from metal monofilament and PES multifilament yarn		3 plied yarn formed from metal monofilament and 2 cotton yarns	
Composition	PES (82 %), steel (18 %)	PES (67 %)	Steel filament (33 %)	Cu filament coated with Ag (50 %)	Cotton (50 %)
Single yarn linear density, tex	30	20	10	10	10
Total linear density of plied yarn, tex	30	33		20	
Yarn thickness, mm	0.13	0.12		0.09	
Fiber diameter (d), μm	-PES fiber, d=20; -steel staple fiber, d=10	-PES fiber, d=20; -steel fiber, d=30		-cotton fibers; -Cu fiber coated with Ag, d=30	
Resistance per 10 cm	2 kΩ	3.8 Ω		84.9 Ω	

2.2. Preparation of knitted samples

Knitted samples were developed in single jersey and 1x1 rib knitting pattern with manual flat knitting machines of gauge 7. The gauge number refers to the number of needles per inch. The larger the gauge number, the finer the knitting.

After the first attempts of sample creation, it was decided to add cotton (co) yarn for such aspects as, firstly, providing a smooth knitting process because the metal-contained yarns were quite fine and had a tendency to break during the first processing attempts, and secondly, to obtain soft-touch textiles for potential use near the skin. Totally 14 samples (seven single jersey and seven 1x1 rib pattern) were made of above mentioned yarns (A-co, B-co, C-co) and their combinations (A-B, A-C-co, B-C, and A-B-C). Some magnified samples are presented in Figure 1.

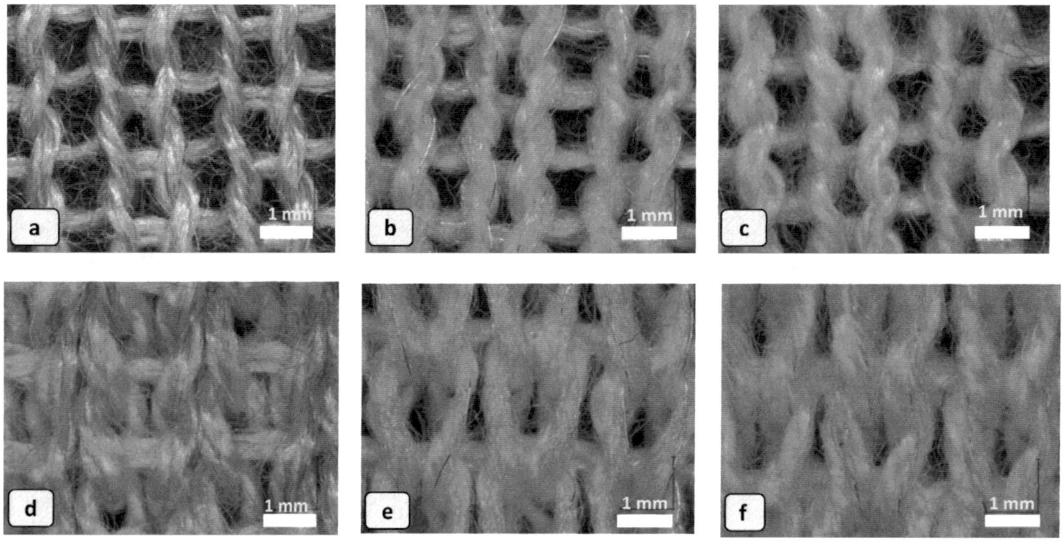

Figure 1. Knitted textile samples (magnification 50 times); single jersey samples: a – yarn A doubled with cotton yarn, b – yarn B doubled with cotton yarn, c – yarn C doubled with cotton yarn; 1x1 rib samples: d - yarn A doubled with cotton yarn, e – yarn B doubled with cotton yarn, f – yarn C doubled with cotton yarn

2.3. Determination of characteristics of knitted samples

The thickness of knitted samples was measured according to the standard LVS EN ISO 5084:2001 "Textiles - Determination of thickness of textiles and textile products" [18]. The stitch length and number of stitches per unit length were estimated to the standards LVS EN 14970:2006 "Textiles – Knitted fabrics – Determination of stitch length and yarn linear density in weft knitted fabrics" and LVS EN 14971:2006 "Textiles – Knitted fabrics – Determination of number of stitches per unit length and unit area" accordingly [19, 20].

2.4. Determination of wave attenuation

A test set of two horn antennas *Rohde & Schwarz HF 906* and a vector network analyzer (VNA) *Anritsu MS2024B* was used. The distance between the source of signal radiation and the signal receiver was 1.5 m (see Figure 2). The selected antennas and the VNA device cover different frequency ranges; therefore, measurements were performed in the range of 2.4 GHz to 2.5 GHz. The scattering parameters s11 (input port reflection) and s12 (reverse gain) were measured from antenna 1 to antenna receiver 2. At the beginning, the reference measurement was done without the sample, then the knitted sample was put between the antennas, but closer to antenna 1. The attenuation of the s12 parameter was measured by VNA in dB, and the difference between the s12 parameter value with and without a sample was calculated in relative values.

Shielding efficiency (SE) is characterized as a ratio of transmitted wave intensity (I_{TR}) to incident intensity (I_{INC}), referring to (1).

$$SE = \frac{I_{TR}}{I_{INC}} \tag{1}$$

Attenuation (A) of transmitted electromagnetic waves is calculated as:

$$A\,(\%) = (1 - SE) \cdot 100 \tag{2}$$

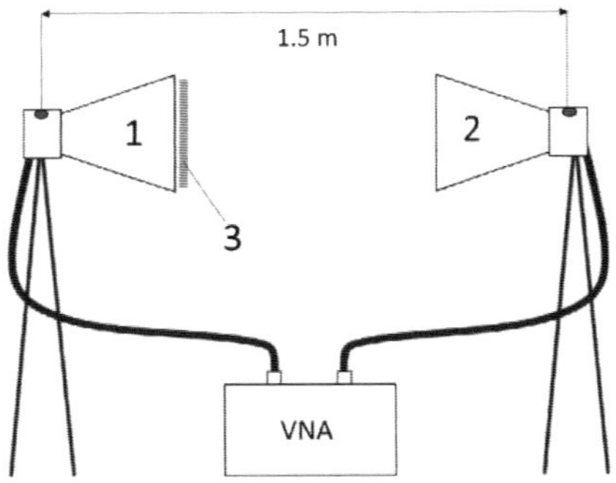

Figure 2. Test set of two horn antennas 1 and 2, VNA, and knitted sample -3

3. Results and Discussion

3.1. Characteristics of knitted textile samples

To characterize the knitting structure of textiles, the amount of wales and courses per cm, stitch length, and thickness were determined (Table 2.). The number of courses per cm differs little; single jersey samples have 1.5 or 2 courses more than the 1x1 rib samples. But the amount of wales per cm of 1x1 rib samples is almost twice as much as for single jersey samples. As well, the rib samples are in average 70% thicker than single jersey ones, which is explained by the spatial structure of the pattern type.

Table 2. Characteristics of knitted samples

Parameters	Knitting pattern	A-co	B-co	C-co	A-B	A-C-co	B-C	A-B-C
Courses per cm	Single jersey	7	7	7	8	7.5	7.5	8
Wales per cm		5	5.5	5	5	5	5	5
Stitch length, mm		7.26	5.82	6.96	6.00	6.89	6.58	5.99
Thickness, mm		1.01	1.08	1.33	0.93	1.18	1.19	1.06
Courses per cm	1x1 rib	6.5	5	5	6	6	5.5	6.5
Wales per cm		9	9	9	9	10	10	9
Stitch length, mm		7.94	8.67	8.45	7.46	7.37	7.41	7.90
Thickness, mm		1.62	1.76	2.12	1.74	2.03	2.01	1.85

In Table 3, the mass per square meter of knitted samples and the concentration of metal fibers are presented. Comparing the mass (g) per square meter of knitted samples accordingly, it is seen that 1x1 rib samples are, on average, 67% heavier than single jersey samples, except for the samples of A-B-C yarns, where the rib sample is twice as heavy as the single jersey one. In the single jersey pattern group, the lightest sample (127 g/m^2) is made of B-C yarns, but the heaviest sample (250 g/m^2), is consisting of C yarn doubled with two cotton yarns. In 1x1 rib pattern group, the lightest sample is B-C (219 g/m^2), but the heaviest sample is A-C-co (393 g/m^2). Comparing the mass per square meter of metal fibers, the rib samples are 63% heavier than single jersey samples. The sample A-co contains the least mass of metal in both pattern groups; in single jersey group it is 15 g/m^2, in the rib group it is 24 g/m^2. But the heaviest mass of metal are for the samples of A-B-C yarns, 64 g/m^2 and 105 g/m^2, respectively.

Table 3. Mass per unit area of knitted samples and metal fibers

Parameters	Knitting pattern	A-co	B-co	C-co	A-B	A-C-co	B-C	A-B-C
Mass per unit area, g/m^2	Single jersey	195	176	250	157	234	127	190
Mass per unit area of metal fibers, g/m^2		15	26	24	40	39	50	64
Mass per unit area, g/m^2	1x1 rib	321	299	388	268	393	220	386
Mass per unit area of metal fibers, g/m^2		24	43	37	68	61	81	105

3.2. Results of wave attenuation in knitted samples

The transmittance of EM waves was characterized by the calculation of it attenuation, according to Equation (2). In Figure 3 and Figure 4, the values of the attenuation are given for single jersey and 1x1 rib samples positioned vertically to the radiation source at frequency of 2.4 – 2.5 GHz.

The highest attenuation (above 99%) of single jersey samples is observed for samples consisting of all three types (A-B-C) of yarns and combination of A, C, and cotton yarn. On the one hand, it can be explained by the synergy of metal components (steel staple, steel, and Ag-coated Cu filament) of the yarns. On the other hand, it can be reasoned by the fact of using three yarns at once in the knitting process, and in a result, the structure of knitting is denser and more stable, which enables high attenuation. Relatively low attenuation is for sample C-co and B-co.

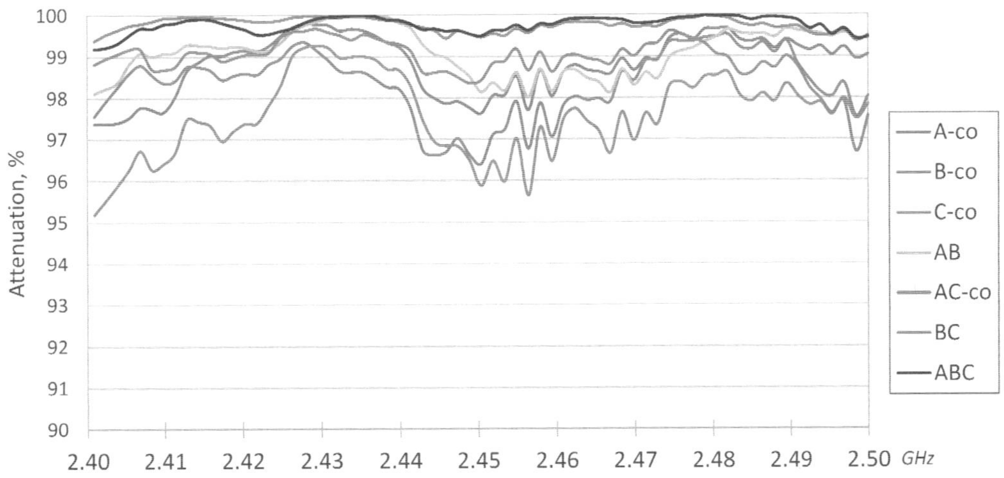

Figure 3. Attenuation of single jersey samples (positioned vertically), at 2.4-2.5 GHz frequency

The attenuation level of 1x1 rib samples is lower (under 95%) comparing to single jersey ones. It can be explained due to the knitting pattern, as for the rib pattern samples, yarn consumption is bigger, due to rib samples are 67% heavier (Table 3) and thicker than single jersey samples; consequently, the electrical resistance of longer yarn increases and the attenuation level decreases, because higher conductivity of the yarn, leads to increased shielding effectiveness. The most promising/best result from this group of samples is shown with the sample A-B, containing steel staple and steel filament, which attenuates the signal above the level of 95 %. Similar attenuation tendency is observed for samples A-co, AC-co, BC, and ABC, the attenuation is between 85-95%. The lowest results (as in single jersey sample group) are for C-co, attenuation between 55-72%.

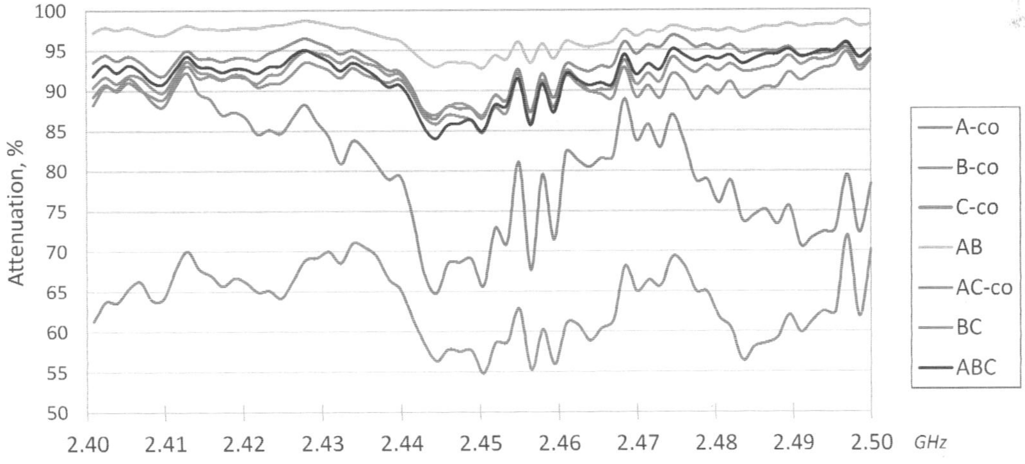

Figure 4. Attenuation of 1x1 rib samples (positioned vertically), at 2.4-2.5 GHz frequency

Conclusions

In this research, the dependence of the attenuation of electromagnetic waves by samples on the type and combination of metal-contained yarns and knitting patterns was analyzed. It was clarified that knitted samples (single and 1x1 rib) increase the attenuation level. Single jersey samples attenuate the radiation signal of 2.4.-2.5 GHz better than 1x1 rib-knitting samples of the same yarn combination due to the fact that the mass of metal fibers per square meter of single jersey is 67% less than for rib ones. Consequently, the less consumption of metal contained yarns, the better conductivity and higher attenuation level.

References

[1] Genuis S J 2007 Fielding a current idea: Exploring the public health impact of electromagnetic radiation *Public Health (London)* vol **122(2)** pp 113-124 DOI: 10.1016/j.puhe.2007.04.008.
[2] Röösli M, Moser M, Baldinin Y, Meier M, Braun-Fahrländer C 2004 Symptoms of ill health ascribed to electromagnetic field exposure – a questionnaire survey, *International Journal of Hygiene and Environmental Health* vol **207(2)** pp 141-150
[3] Pacchierotti F *et al* 2021 Effects of Radiofrequency Electromagnetic Field (RF-EMF) exposure on male fertility and pregnancy and birth outcomes: Protocols for a systematic review of experimental studies in non-human mammals and in human sperm exposed in vitro *Environment International*, vol **157** 106806 ISSN 01604120 (https://www.sciencedirect.com/science/article/pii/S0160412021004311)
[4] Schuermann D, Mevissen M 2021 Manmade Electromagnetic Fields and Oxidative Stress —Biological Effects and Consequences for Health *Int. J. Mol. Sci.* vol **22**, 3772 https://doi.org/10.3390/ijms22073772
[5] Pall M L 2018 Wi-Fi is an important threat to human health *Environmental Research* vol **164** pp 405-416 ISSN 0013-9351, https://doi.org/10.1016/j.envres.2018.01.035.
[6] GUIDELINES FOR LIMITING EXPOSURE TO ELECTROMAGNETIC FIELDS (100 kHz to 300 GHz) International Commission on Non-Ionizing Radiation Protection (ICNIRP) DOI: 10.1097/HP.0000000000001210
[7] Latvian Association of Free Trade Unions 2010 Occupational hygiene *Ministry of Welfare of Latvia. Riga:*/Latvijas brīvo arodbiedrību asociācija. Darba higiēna. Rīga: Latvijas labklājības ministrija, pp. 137-139
[8] 3M Electronics. [Online]. Available: http://solutions.3m.com/wps/portal/3M/en_US/Electronics_NA/Electronics/Products/Product_Catalog.
[9] Less EMF. [Online]. Available: http://www.lessemf.com/fabric.html#1232
[10] Lorix Ltd. [Online]. [viewed 6 May 2013]. Available from: http://lorix-en.shp.hu/hpc/web.php?a=lorix-en&o=non_woven_metallized_textiles_r2AQ
[11] Marburg Technic. [Online]. Available: http://www.martech-systems.de/no_cache/en/products/shop/pl/single-view.html?tt_products%5BbackPID%5D=199&tt_products%5Bproduct%5D=715&cHash=cea0ee82681d225b0ae9d05b8c0a78bb.
[12] SGL Group. [Online]. Available: https://www.sglgroup.com/cms/international/products/product-groups/cf/carbon-fibers-for-intake-fittings/index.html?__locale=en.
[13] Swiss Shield Textile. [Online]. Available: https://www.swiss-shield.ch/index.php?id=25
[14] Ceken F *et al* 2012 Electromagnetic Shielding Properties of Plain Knitted Fabrics Containing Conductive Yarns *Journal of Engineered Fibers and Fabrics* vol **7**
[15] Pušic T, Šaravanja B, Malaric K 2021 Electromagnetic Shielding Properties of Knitted Fabric Made from Polyamide Threads Coated with Silver *Materials*, vol **14** 1281
[16] Palanisamy S *et al* 2024 The effect of tensile deformation of knitted fabrics on their electromagnetic shielding ability, electrical resistance, and porosity *Journal of Industrial Textile* vol **54**
[17] Electronics Notes, Wi-Fi Channels, Frequencies, Bands & Bandwidths. [Online]. Available: https://www.electronics-notes.com/articles/connectivity/wifi-ieee-802-11/channels-frequencies-bands-bandwidth.php
[18] LVS, Standarti. LVS EN ISO 5084:2001Textiles - Determination of thickness of textiles and textile products, https://www.lvs.lv/lv/products/3921
[19] LVS, Standarti, LVS EN 14971:2006 Textiles - Knitted fabrics - Determination of number of stitches per unit length and unit area, https://www.lvs.lv/lv/products/22931
[20] LVS, Standarti, LVS EN 14970:2006 Textiles - Knitted fabrics - Determination of stitch length and yarn linear density in weft knitted fabrics, https://www.lvs.lv/lv/products/19660

IOP Conf. Series: Materials Science and Engineering **1320** (2024) 012005 doi:10.1088/1757-899X/1320/1/012005

Tri-band planar antenna for a small robotic vehicle

M Richterova[1*], J Virgala[1], M Popela[2] and P Kuzel[1]

[1] Department of Communication Technologies, Electronic Warfare and Radiolocation. University of Defence, Kounicova 65, 662 10, Brno, Czech Republic
[2] Faculty of Electrical Engineering and Communication, Brno University of Technology, 601 90 Brno, Czech Republic

*E-mail: marie.richterova@unob.cz

Abstract. This paper describes the design, analysis of properties and fabrication of a tri-band antenna for a small robotic vehicle. The E-shaped tri-band antenna was chosen for its simplicity, practicality and universality. The design of the tri-band antenna and the simulation of its properties were implemented in the Matlab programming environment using the Antenna Toolbox extension. The designed tri-band antenna was fabricated, its properties were measured and the measured results are compared with the simulations for subsequent analysis.

1. Introduction

Antennas have found use in many technical applications, but the dominant area of their deployment is radio communications. To fulfil the essence of mobile communications systems, it was necessary to achieve antenna sizes on a mobile robotic vehicle such that the antenna could be integrated into the device and capable of working in multiple frequency bands.

This requirement also translated into the design of antennas that are suitable for mobile robotic vehicles. Another factor that has influenced the design of antennas is the multitude of services that contemporary mobile robotic vehicles provide. To mention a few, image, video and data transmission or GPS positioning, each of these applications needs its own antenna.

Various small autonomous robotic vehicles are used for terrain monitoring, where they are able to identify various terrain obstacles such as terrain faults or sinkholes, or various types of objects such as explosives, boxes or luggage. A way to reduce the number of antennas on a mobile small robotic vehicle may be to utilize multiband antennas. One type of antennas that meet the demands placed on mobile communications are microstrip antennas [1].

1.1 Microstrip antennas

The requirements for a compact antenna with small size, low weight, easy installation on the object of interest and cheap manufacturing led to the development of microstrip antennas during the 1970s. Due to its characteristics, the microstrip antenna has found use in both military and commercial applications. To mention a few, deployments in aerospace, rocketry, space industry, mobile phones, GPS receivers, and laptops [2], [3], [4], [5].

Due to the frequency trends towards higher frequencies in the MHz and GHz order, there have also been substantial changes in the size and shapes of microstrip antennas. Gradual developments in communication technology are moving from the original frequency bands used to higher frequencies. This is a practical way to deal with frequency band saturation and allows smaller antennas to be designed.

IOP Conf. Series: Materials Science and Engineering **1320** (2024) 012005 doi:10.1088/1757-899X/1320/1/012005

Since the antenna dimensions are derived from the wavelength, it follows that the size of antennas cannot be continuously reduced, as is the case with integrated circuits. Therefore, the limit for antenna size is not the technology used, but the laws of physics. However, the trend towards higher frequencies has made it possible to construct antennas of practical dimensions, suitable not only for the various mobile devices with which mobile communications can be realised. The microstrip antennas are thus used contemporary in the UHF and SHF bands [4], [5], [6].

However, some conventional microstrip antennas are still too oversized for some applications. The limiting factor for miniaturization of wireless devices is usually the antenna. As mentioned, there are limits to the size of antennas, but there is still some space for downsizing microstrip antennas.

There are some methods available to design an antenna with small size. These methods include, for example, cutting various slots in the radiating element of the antenna, flexing the entire microstrip antenna including the substrate and ground plane, or flexing only the radiating element. More details on these methods are presented in [7].

2. Analysis of tri-band antenna properties

The tri-band antennas are small planar antennas capable of operating in the 900 MHz, 1800 MHz, 2.4 GHz and higher frequency bands. Due to their small size, these antennas are suitable as antennas for small robotic autonomous vehicles. If the autonomous robotic vehicle is to move in open terrain, it is expected that radio communication over several hundred meters will be required.

A frequency of 900 MHz is the optimum operating frequency due to the dielectric properties of the earth's surface at that frequency. Conversely, if the small robotic rover will be moving in a building or in an urban area, then operating frequencies of 1800 MHz and above are suitable for radio communication over several tens of meters [8], [9]. The size and weight of the small robotic vehicle are limited, so it is advisable to take advantage of the small size and optimal characteristics of the tri-band planar antenna. In the following section, the design and characteristics of the E-shaped planar antenna will be described.

2.1 Design of the E-shaped planar antenna

The design of a single-band microstrip antenna uses the well-known relations available in [10] to determine the width W and length L of the radiating element. In the design of a tri-band antenna, the problem is more complicated because, since the proposed antenna must be capable of operating in several frequency bands, it is necessary to view the radiating element as if it were several separate radiating elements.

The E-shaped microstrip planar antenna is designed for 900MHz, 1800 MHz and 3700 MHz. The E-shaped antenna will be simulated in Matlab programming environment using Antenna Toolbox. In all simulations, FR4 substrate with permittivity ε_r = 4.8 will be used as dielectric material.

The basic dimensions of the E-shaped microstrip antenna for all three resonance frequencies are L1 = 76.93 mm, L2 = 37.9 mm and L3 = 18.29 mm.

The lengths of all three antenna arms are determined, now we need to determine their width, the dimensions of the base and the overall dimensions of the antenna including the ground plane on the FR4 substrate. The length of the base should be the same as the length of the arm for the lowest resonant frequency, therefore we choose the length L1 (for 900 MHz).

The width of the base and the individual arms is set as 20 % of the length L1, i.e. 15.34 mm. The radiating element so designed is placed on the centre of the substrate, the width and length of which are obtained by increasing the width and length of the radiating structure, respectively, by 10 %.

The basic antenna dimensions calculated in this way are presented in Table 1. These values are the input values for achieving the desired antenna characteristics at resonant frequencies. From the basic dimensions, we create a model of the antenna in Matlab and analyze the radiation and impedance properties. Based on the simulation results, the antenna dimensions are then optimised until the desired radiation characteristics are achieved at the selected resonance frequencies.

Table 1. Basic antenna sizes (mm) and their optimization.

	Arm 1		Arm 2		Arm 3	
	Length	Width	Length	Width	Length	Width
Basic size	76.93	15.34	37.9	15.34	18.29	15.34
Optimal size	72.2	28.6	34.6	19.6	21.8	16.6
	Base		Substrate		Ground	
	Length	Width	Length	Width	Length	Width
Basic size	76.93	15.34	84.62	84.62	84.62	84.62
Optimal size	115	11.6	126.52	83.8	101.2	71.2
	SMA connector					
	X coordinate	Y coordinate				
Basic position	-33.3	0				
Optimal position	-5	0				

The proposed E-shaped planar antenna is shown in Figure 1.

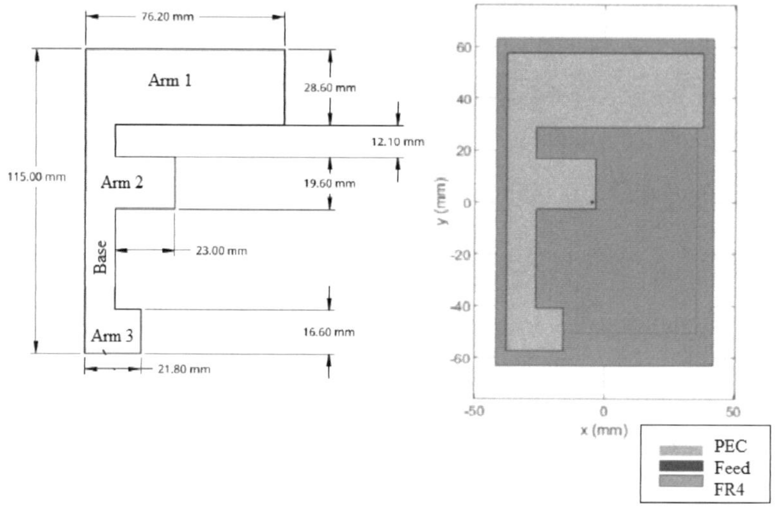

Figure 1. The E-shaped planar antenna a) quoted antenna dimensions b) Matlab representation of the antenna on the substrate.

It is shown that the lengths of the individual arms are determined with a high degree of precision and only minimal dimensional changes occur. However, the width of the arms has to be modified to a greater extent, as it has only been estimated as 20 % of the length of the longest arm. The greatest change is seen in the dimensions of the base, which increase by up to several centimetres. The new dimensions of the radiating element E will then be reflected in the overall dimensions of the PEC base. Furthermore, we will reduce the ground plane, which in the original model occupies the entire substrate area.

A change that is not presented in Table 1 is the shift of the centre arm (arm 2) by 7 mm towards the arm 1. This means that the spacing between the arms is not uniform in the final design. The last change is the shift of the SMA feed connector close to the centre of the substrate.

Thus, by carefully adjusting the dimensions based on the results of the analysis of each step, we achieve the presented values. This final E-shaped antenna variant with resonance frequencies fr1 = 899 MHz, fr2 = 1 784 MHz and fr3 = 3 776 MHz is close to the frequencies f1 (900 MHz), f2 (1 800 MHz) and f3 (3 700 MHz) determined at the beginning of the simulation.

2.2 Analysis of the E-shaped planar antenna

In order to determine the impedance and radiation pattern of the E-shaped planar antenna, we must first analyze the current distribution on the antenna surface. The current distribution gives us information that the feed point is correctly chosen and that the designed antenna will be able to radiate. The Figures 2 to 4 show the current distribution for the resonance frequencies fr1, fr2 and fr3.

IOP Conf. Series: Materials Science and Engineering **1320** (2024) 012005 doi:10.1088/1757-899X/1320/1/012005

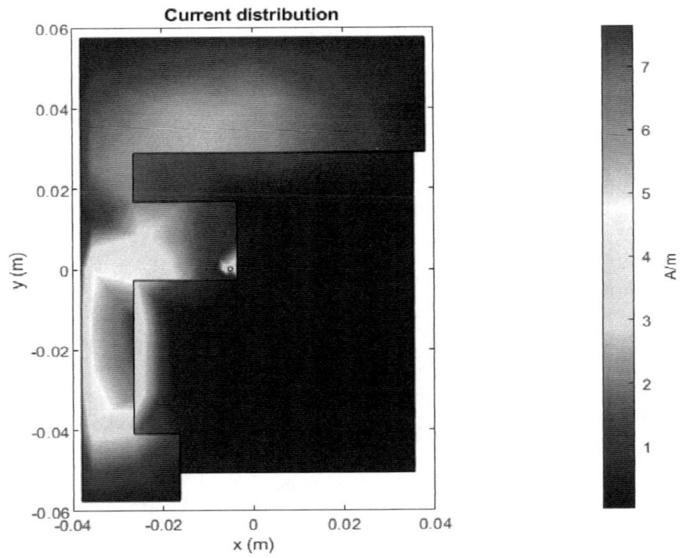

Figure 2. Simulation of the the surface current distribution for frequency fr1 = 899 MHz.

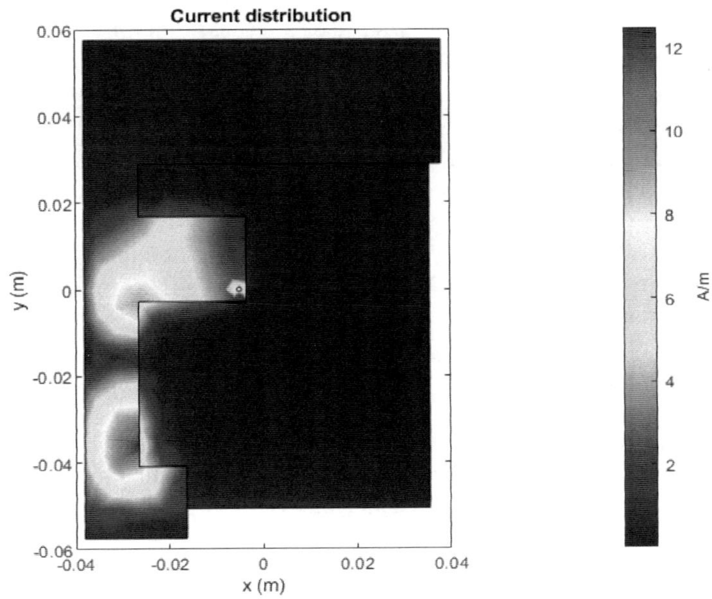

Figure 3. Simulation of the the surface current distribution for frequency fr2 = 1784 MHz.

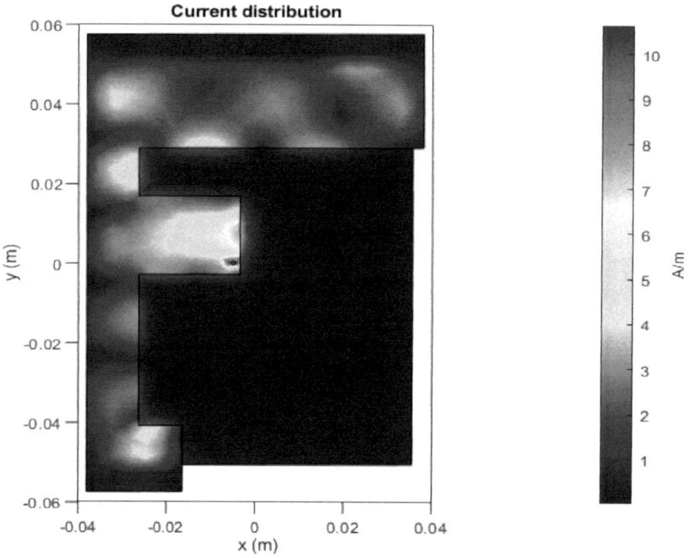

Figure 4. Simulation of the surface current distribution for frequency fr3 = 3 776 MHz.

The E-shaped planar antenna is designed with the idea that one specific arm is designated for each band and the current distribution for that frequency will correspond to that. Despite these assumptions, the simulated current distribution is different. For resonance fr1, some current values can be observed on arm one, however, the dominant area of current concentration is the part of the base between arms two and three and the feed connector area.

For the second resonance fr2, the current is present on arm two as expected, but reaches its maximum values in the part of the base near the third arm. The uniform distribution of the current over the surface of the radiating element and its maximum around the feed connector is the condition corresponding to the frequency fr3.

The impedance of the E-shaped planar antenna was calculated in the frequency band around the resonance frequencies fr1, fr2 and fr3 in order to determine the exact resonance frequencies and to investigate any anomalies in the impedance charts (see Figure 5).

The power radiation patterns for vertical polarization, horizontal polarization and resonance frequencies fr1, fr2 and fr3 are presented in Figures 6 and 7.

In the imaging plane, the directional characteristic for frequency fr1 resembles the behaviour of a dipole and produces two opposite equal-sized lobes. As the frequency increases, the pattern changes, and for frequency fr2 the main lobe is seen to increase (from -50 dB to -30 dB) at the expense of the rear lobe, which does not disappear completely. For the third resonant frequency, two side lobes are added to the main and back lobes.

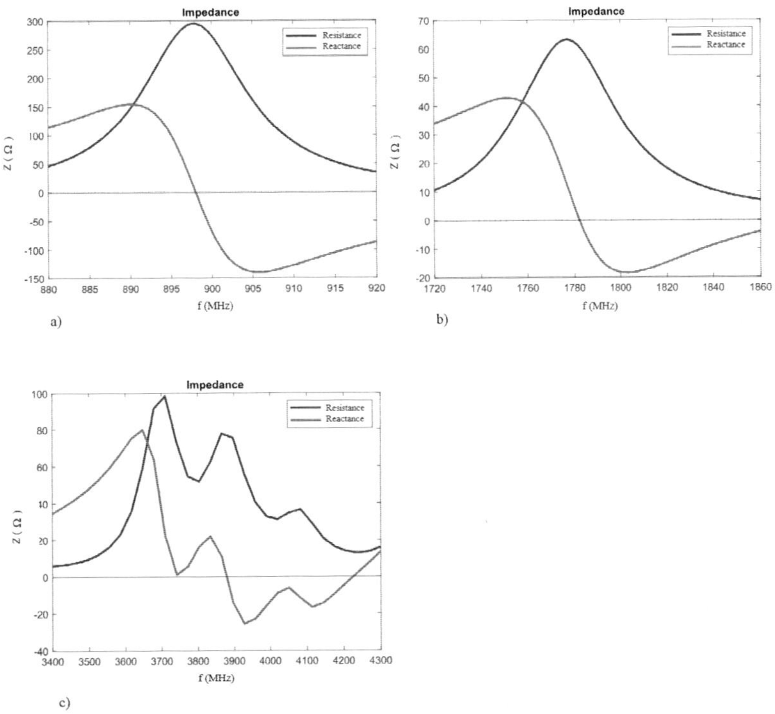

Figure 5. Simulated impedance of an E-shaped planar antenna for frequencies a) fr1 = 899 MHz; b) fr2 = 1 784 MHz; c) fr3 = 3 776 MHz.

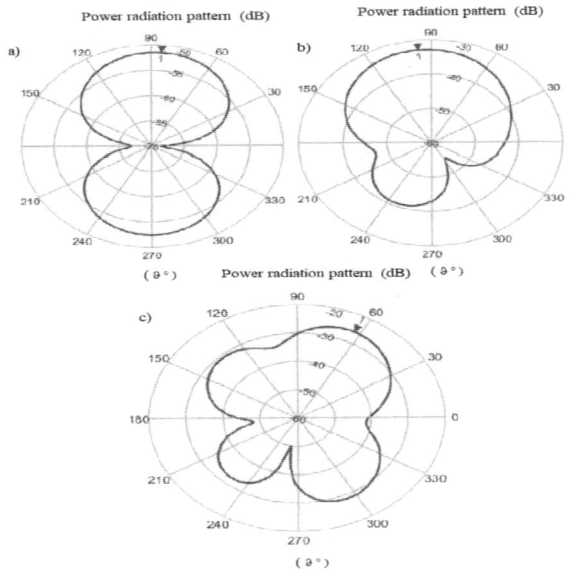

Figure 6. Simulated power radiation characteristics for vertical antenna polarisation for resonant frequencies (a) fr = 899 MHz; (b) fr2 = 1 784 MHz; (c) fr3 = 3 776 MHz.

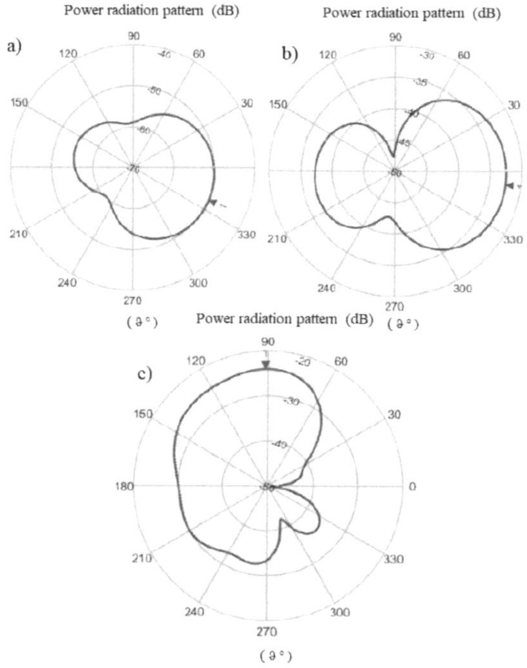

Figure 7. Simulated power radiation characteristics for horizontal antenna polarisation for resonant frequencies (a) fr = 899 MHz; (b) fr2 = 1 784 MHz; (c) fr3 = 3 776 MHz.

Simulations of the power radiation patterns confirmed the assumptions that the proposed E-shaped planar antenna is capable of operating in three frequency bands. The directional properties of the proposed E-shaped planar antenna and its size will satisfy the requirements for the application of this type of antenna on a small mobile robotic vehicle.

The proposed E-shaped planar antenna was fabricated using photolithography technology. For financial reasons, the FR4 substrate is chosen, but it is not very suitable for applications above 1 GHz. The aim was to verify the function of the antenna and then to place it on a small mobile robotic vehicle.

3. Antenna performance measurements and results

In order to compare the radiation patterns of the simulated E-shaped planar antenna and the antenna fabricated on the substrate, it is necessary to perform measurements in an anechoic chamber to eliminate the negative influence of the surrounding environment on the experiment.

Measurements of the power radiation patterns for the vertical and horizontal polarization of the fabricated tri-band antenna were performed in a anechoic chamber at the Department of Communication Technology, Electronic Warfare and Radiolocation at the University of Defence.

The measured power radiation patterns for vertical polarization, horizontal polarization and resonance frequencies fr1, fr2 and fr3 are shown in Figures 8 and 9.

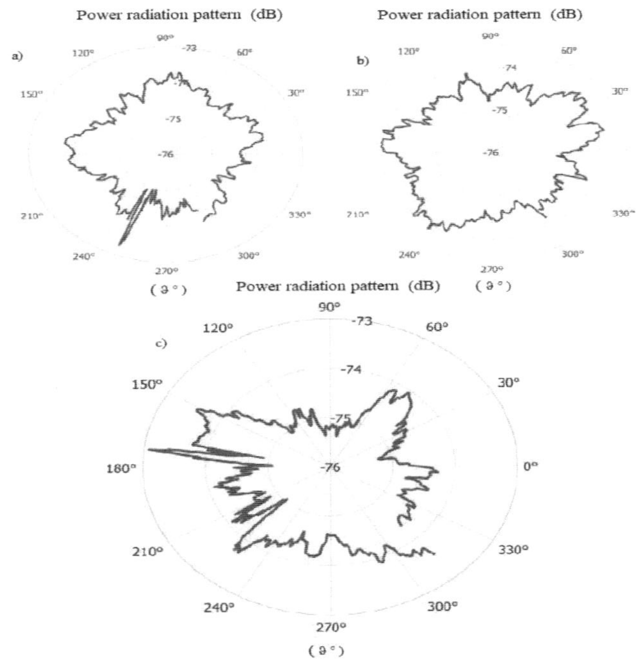

Figure 8. Measured power radiation characteristics for vertical antenna polarisation for resonant frequencies (a) fr = 899 MHz; (b) fr2 = 1 784 MHz; (c) fr3 = 3 776 MHz.

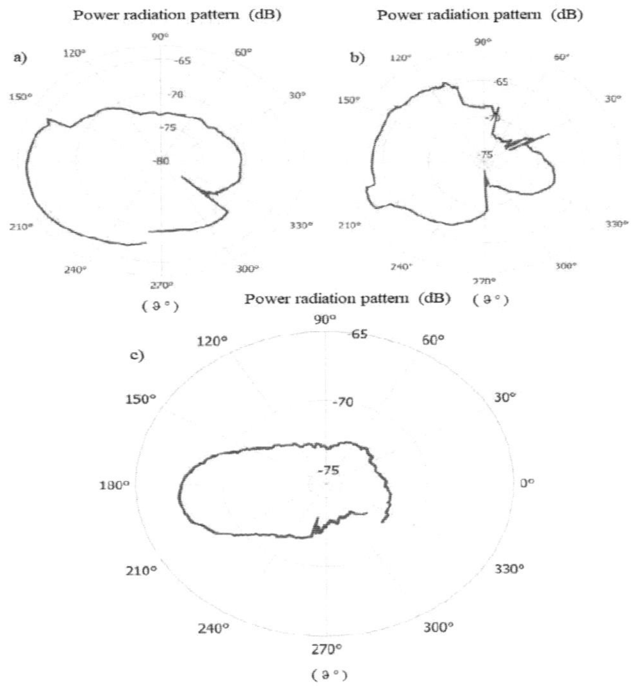

Figure 9. Measured power radiation characteristics for horizontal antenna polarisation for resonant frequencies (a) fr = 899 MHz; (b) fr2 = 1 784 MHz; (c) fr3 = 3 776 MHz.

The measured power radiation patterns proved that the designed and fabricated three-band planar antenna is functional. Experiments have demonstrated that the FR4 substrate is not suitable for frequencies higher than 1 GHz, therefore the measured power radiation patterns only partially correspond to the simulated power radiation patterns.

4. Conclusion

The obtained results show that tri-band planar antenna is suitable for applications where miniature dimensions and low device weight are required. The designed and realized E-shaped planar antenna can be used for military applications as well as for civilian applications, e.g. for small robotics vehicles and for 5G communication technologies.

Further research on the designed E-shaped planar antenna will focus on the selection of a substrate with optimal properties for frequencies higher than 1 GHz. Experiments have shown that it is also necessary to design and implement matching elements to achieve optimal impedance ratios between the antenna and the control unit of a small robotic vehicle.

Acknowledgments

The work presented in this article was supported by the Czech Republic Ministry of Defence – the University of Defence development program "AIROPS".

References

[1] Balanis C A 2016 Antenna Theory Analysis and Design (Hoboken, New Jersey: John Wiley & Sons)

[2] Hussain M, Jarchavi S M R, Naqvi S I, Gulzar U, Khan S, Alibakhshikenari M and Huynen I 2021 Design and Fabrication of a Printed Tri-Band Antenna for 5G Applications Operating across Ka-, and V-Band Spectrums *Electronics* **10** 2674

[3] Thanamalapong W, Phongcharoenpanich C and Lamultree S 2020 A Tri-Band Antenna for 2.4/5 GHz WLAN and Ku-Band Applications *2020 17th International Conference on Electrical Engineering/Electronics, Computer, Telecommunications and Information Technology* (ECTI-CON) pp 80-83

[4] Gao X, He Y, Zhang L, Zhang L, Li W, Wong S-W and Zeng Z 2021 A Tri-band Dual-Polarized Base Station Antenna for Sub-6 GHz 5G Communication Systems *2021 International Conference on Microwave and Milimeter Wave Technology* (ICMMT) pp 1-3

[5] Zhang Y, Jiao Y-C, Yang M-M and Zhao B-L 2019 A tri-band notched Ultra-wideband MIMO Antenna *2019 International Symposium on Antennas and Propagation* (ISAP) pp 1-3

[6] Chen Z N and Chia M Y W 2006 Broadband Planar Antennas: Design and Applications (West Sussex: Wiley)

[7] Stutzman W L and Thiele G A 2013 Antenna theory and design (Hoboken, NJ: Wiley)

[8] Rao P K and Mishra R 2018 A Disc Shape Monopole Antenna with Tri- Band-Notched 2.55/3.5/5.5 GHz Characteristics *2018 IEEE International Students' Conference on Electrical, Electronics and Computer Science* pp 1-6

[9] Hernando J and Peréz-Fontán F 1999 Introduction to Mobile Communications Engineering (Artech House Publishers)

[10] Balanis C A 2008 Modern antenna handbook (Hoboken, New Jersey: John Wiley & Sons)

IOP Conf. Series: Materials Science and Engineering **1320** (2024) 012006 doi:10.1088/1757-899X/1320/1/012006

Design and Simulation of Three-Band Microstrip Resonator Bandpass Filters for Wireless Communication Systems

Sergiu Andreica, Calin Munteanu, Claudia Constantinescu, Claudia Pacurar, Adina Giurgiuman, Marian Gliga and Sabin Dusa

Department of Electrotechnics and Measurements, Technical University of Cluj-Napoca, Cluj-Napoca, Romania

E-mail: Sergiu.Andreica@ethm.utcluj.ro

Abstract. This research focuses on the design, simulation, and practical implementation of three-band microstrip resonator bandpass filters for wireless communication systems. Key parameters targeted in optimizing these filters include coupling coefficient, external quality factor, and fractional bandwidth, among others. Using ANSYS HFSS software, the electromagnetic behavior of the filters was analyzed to enhance their design. The proposed filters were fabricated on RT/Duroid 5880 substrate and experimentally validated using a vector network analyzer. Experimental results were compared with simulations, demonstrating satisfactory performance at specific frequencies relevant to Wi-Fi, 5G, and radar applications. Additionally, a comparative study with other similar filters from the literature was conducted to evaluate their performance.

1. Introduction

The multi-band filter has been of great interest in recent decades for modern wireless communication systems. The microstrip multi-band filter can be easily fabricated and integrated on a dielectric substrate, offering significant design flexibility when precisely measuring the width, length, and thickness of the microstrip.

Filters with more bands have been designed by many authors in the specialized literature, but the complexity of circuit integration, design methods, and achieving the desired application design are currently the most significant challenges for miniaturized band-pass filters (BPFs). In RF(radio frequency) and microwave communication devices, wireless communication, and multi-service systems, BPFs are developed with considerable interest. In the front-end frequency section, BPFs play a key role in selecting the desired bandwidths and attenuating unwanted signals [1]-[3].

Filters play a crucial role in many applications, leading to increasingly stringent requirements for higher performance, smaller size, lighter weight, and lower costs. Recent advancements in new materials and manufacturing technologies, including high-temperature superconductors (HTS), low-temperature co-fired ceramics (LTCC), monolithic microwave integrated circuits (MMIC), microelectromechanical systems (MEMS), and microfabrication technology, have stimulated the rapid development of new microstrip and other filters for RF/microwave applications. Meanwhile, advancements in computer-aided design (CAD) tools, such as full-wave electromagnetic (EM) simulators, have revolutionized filter design [4].

The key parameters targeted in the design and optimization of the microstrip resonator BPF filter are the fractional bandwidth, coupling coefficient, and external quality factor described in the relationships below:

$$\text{FBW} = \frac{f_2 - f_1}{f_c} \cdot 100 \quad (1)$$

$$Q_{ext} = \frac{f_c}{BW_{3dB}} \quad (2)$$

$$k_c = \frac{f_2{}^2 - f_1{}^2}{f_2{}^2 + f_1{}^2} \quad (3)$$

the fractional bandwidth (FBW) is an indicator of the proportional width of a frequency range relative to its central frequency. In this context, f_2 and f_1 correspond to the upper and lower bounds of the frequency range, respectively, while f_c denotes the central frequency of the range. [5].

2. Design of Three-Band Microstrip Resonator Filters

Designing a three-band microstrip resonator filter in ANSYS involves several critical steps. First, the performance requirements of the filter must be carefully defined, including operating frequencies, bandwidth, and other key characteristics. Next, the filter geometry is designed, and the material properties are specified in the ANSYS HFSS software. Simulations are then performed to evaluate the filter's behavior and adjust parameters as needed.

The filter structure is parameterized to meet the specifications, followed by the fabrication and testing of a prototype to validate real-world performance. If necessary, the filter is refined based on test results, and the entire process is documented in detail. This design process requires expertise in electromagnetism, microstrip design, and the use of ANSYS HFSS software, as well as access to appropriate hardware resources. Figure 1 shows the schematic of the geometric parameters used in creating the filter and the equivalent circuit of the resonator filter [6].

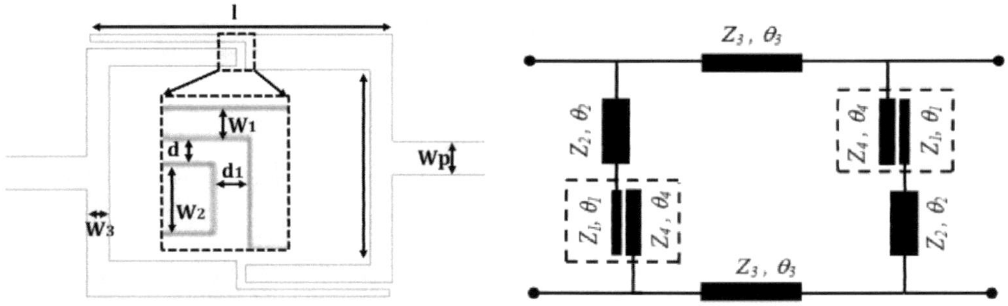

Figure 1. Detailing the geometric dimensions and equivalent circuit for the proposed filter.

In Table 1, the key dimensions of the three-band microstrip bandpass filter are presented, including the resonator length, spacing between resonators, and their width.

The chosen dielectric material, selected for implementing the filter with high performance in RF and microwave domains, is RT/Duroid 5880, with a thickness of 1.27 mm. This material is distinguished by its stable and precise dielectric constant in these frequency ranges, making it suitable for designing and manufacturing microstrip circuits and other RF components.

Moreover, it exhibits low dielectric losses, ensuring excellent signal attenuation performance within the circuit. RT/Duroid 5880 is used in applications requiring precision and stability in the RF and microwave domains.

Table 1. The dimensions of the three-band microstrip bandpass filter.

Parameters and dimensions [mm]	RT/Duroid5880
Resonator length (l)	19.16
Resonator spacing along the Y-axis (d)	0.4
Resonator spacing along the X-axis (d1)	0.58
Microstrip width (W_1)	0.5
Microstrip width (W_2)	1.07
Microstrip width (W_3)	1.42
Microstrip width (W_p)	2.1

3. Numerical modeling of three-band microstrip resonator filters

In the initial phase of developing the three-band microstrip bandpass filter, an electromagnetic behavior analysis was conducted using ANSYS HFSS software. This analysis encompassed two essential aspects: current density and electric field intensity. These parameters were crucial for understanding how radiofrequency (RF) or microwave signals interact with the BPF filter and for evaluating its performance.

Figure 2 illustrates the model implemented in ANSYS HFSS software, comprising several distinct regions. These regions include the air region, dielectric material, ground plane, and resonators.

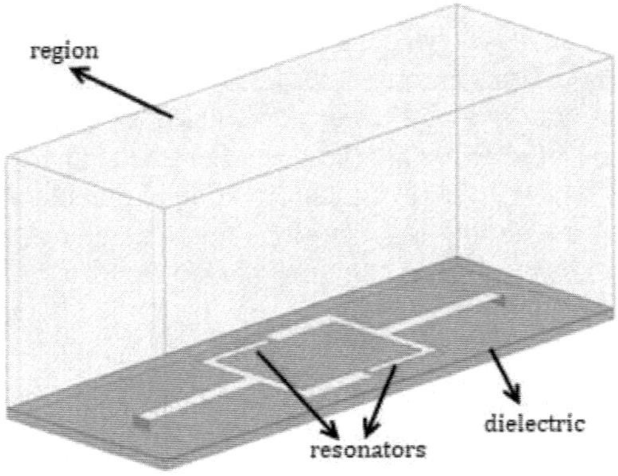

Figure 2. Implementing the three-band frequency filter model in HFSS.

Current density was used to measure and direct the flow of electric charge within the components of the BPF filter, including the resonators and conductors. Through simulations conducted with ANSYS HFSS, we obtained a clear understanding of how electric current distributes and circulates within the filter. This information was particularly useful for evaluating the efficiency of the conductors and identifying areas with high levels of electric current. Additionally, it contributed to detecting potential issues related to heating or energy losses in the filter.

On the other hand, electric field intensity was analyzed to measure the strength and direction of the electric field at various points within the BPF filter. This analysis was important for understanding how RF or microwave signals propagate within the filter and interact with its components. It helped identify areas with high levels of electric field that may require special attention to avoid unwanted phenomena such as interference or significant signal losses.

This electromagnetic behavior analysis constituted an essential first step in the design and development process of the BPF filter, providing a solid foundation for its subsequent optimization to meet specific performance requirements. Figure 3 presents the field magnitudes mentioned above.

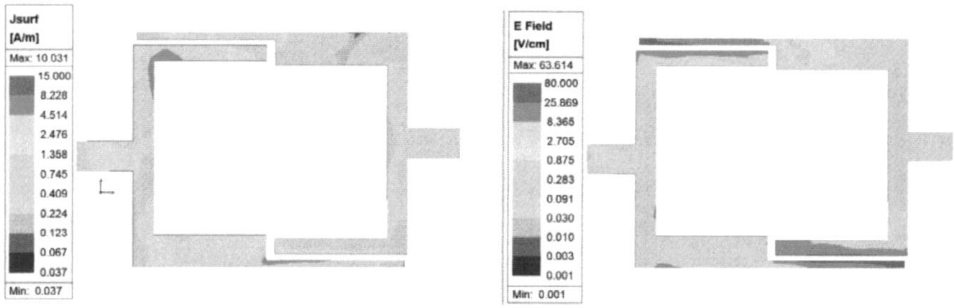

a) b)

Figure 3. Representation of electromagnetic field quantities: a) current density, b) electric field intensity.

4. Optimizing three-band microstrip resonator filters

In choosing the optimal structure for the three-band microstrip bandpass filter, the dielectric plays a crucial role in filter design and significantly influences its performance. The dielectric thickness (g) is critical because it affects resonance frequency, bandwidth, characteristic impedance, dielectric losses, and mechanical compatibility of the filter. Precise control over the dielectric thickness is essential to achieve the desired performance and ensure efficient operation of the filter in wireless communications and other technological domains [7].

Consulting the manufacturer's specifications, the dielectric made from RT/Duroid 5880 has the following standardized thicknesses that were used in this study: 0.25, 0.51, 0.80, 1.27, 1.57. All of these dielectric thicknesses used for the construction of the microstrip bandpass resonator filter were implemented in the ANSYS HFSS software for all three operating bands (2.4 GHz, 3.5 GHz, and 5.2 GHz) and the results obtained are compiled in Table 2.

Table 2. The results obtained from varying the parameter "g".

g [mm]	$f_{1/1'/1''}$ [GHz]	$f_{2/2'/2''}$ [GHz]	$F_{0/0'/0''}$ [GHz]	$BW_{1/2/3}$ [GHz]	$FBW_{1/2/3}$ [%]	$Q_{ext\ 1/2/3}$	$k_{c\ 1/2/3}$
0.25	2.1/3.26/6.65	2.32/3.3/6.71	2.21/3.28/6.68	0.22/0.04/0.06	9.95/1.22/0,9	10/82/111	0.09/0.01/0.008
0.51	2.62/4.48/5.3	3.3/4.83/5.43	2.96/4.65/5.37	0.68/0.35/0.13	22.9/7.52/2.42	4.35/13.3/41.2	0.22/0.81/0.02
0.80	2.5/3.85/5.68	2.58/4.92/5.75	2.93/4.38/5.71	0.86/1.07/0.07	29.3/24.4/1.22	3.4/4.09/81.6	0.28/0.66/0.01
1.27	2.44/3.31/4.43	2.58/3.52/6.2	2.51/3.41/5.31	0.14/0.21/1.77	5.5/6.1/33.3	17.9/1.72/3	0.05/0.53/0.51
1.57	1.91/2.94/6.9	2.84/5.01/7.5	2.37/3.97/7.2	0.93/2.07/0.6	39.1/52.08/8.33	2.55/1.92/12	0.37/0.48/0.08

The parameters analyzed as a function of dielectric thickness include the quality factor and coupling coefficient, and variations of these parameters at the frequency of 2.4 GHz are illustrated in Figure 4.

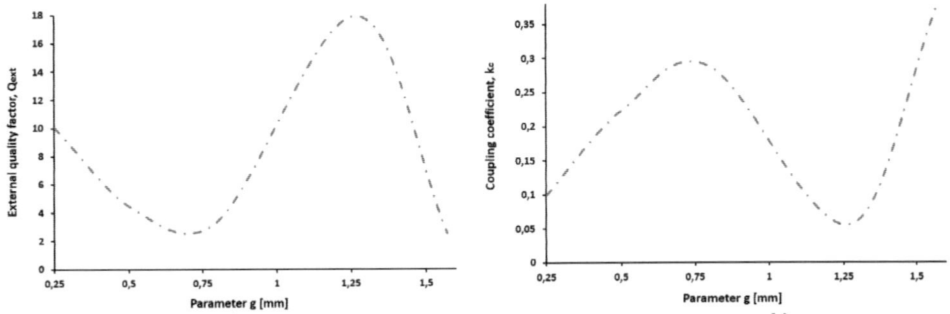

a) b)

Figure 4. Variation of the external quality factor (a) and coupling coefficient (b) as a function of dielectric thickness at 2.4 GHz.

Figure 5 illustrates the variations of the quality factor and coupling coefficient at different dielectric thicknesses for the second frequency band at 3.5 GHz.

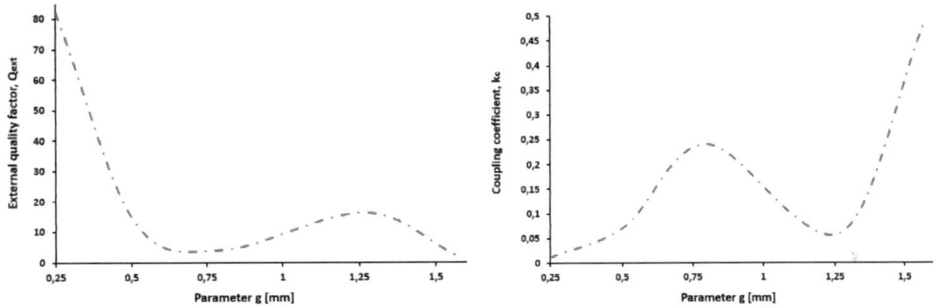

a) b)

Figure 5. Variation of the external quality factor (a) and coupling coefficient (b) as a function of dielectric thickness at 3.5 GHz.

In Figure 6, changes in the quality factor and coupling coefficient are presented as a function of various dielectric thicknesses, with this data analyzed at the third frequency band of 5.2 GHz.

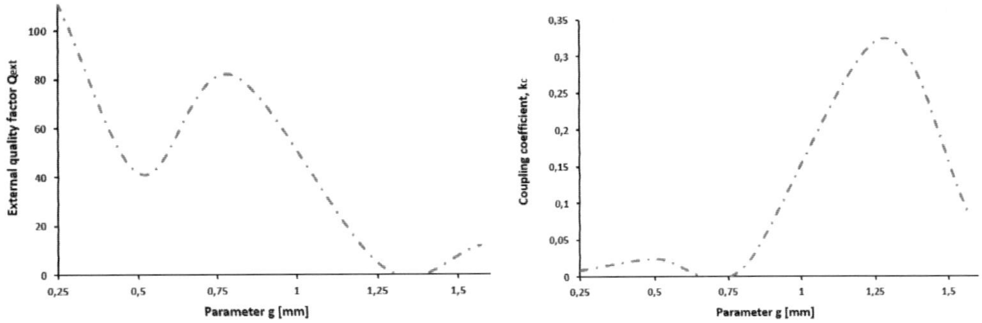

a) b)

Figure 6. Variation of the external quality factor (a) and coupling coefficient (b) as a function of dielectric thickness at 5.2 GHz.

By analyzing the values of the quality factor and coupling coefficient obtained through varying the dielectric thickness to optimize the operation of the microstrip bandpass filter at the respective frequencies of 2.4 GHz, 3.5 GHz, and 5.2 GHz, it was found that the ideal thickness resulting from the parameterization process is 1.27 mm.

Within the conducted analysis, S-parameters were determined to assess the filter performance as shown in Figure 7. One of the key aspects we investigated was the frequency behavior of the filter, and for this purpose, we used S-parameters to evaluate the transmission and reflection of signals in different frequency bands.

We specifically marked the operating bands of the filter at the -3 dB amplitude, which represents a common reference in filter characterization. The microstrip bandpass resonator filter operates at frequencies of 2.4 GHz, 3.5 GHz, and 5.2 GHz, as can be observed in Figure 5.10. At this amplitude, the transmitted or reflected signal in the filter is reduced to approximately 50% of its maximum value, thereby specifying the upper and lower boundaries of the filter's passbands.

This approach allowed for a detailed understanding of the filter's frequency behavior and precise identification of its passbands, which is essential for evaluating its performance in the context of a specific application. This data was subsequently used to validate and adjust the filter design to ensure compliance with performance requirements.

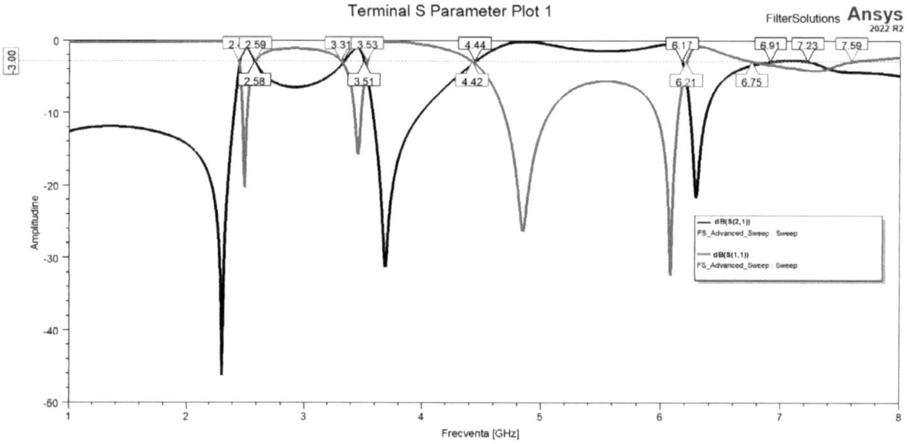

Figure 7. S-parameters of the three-band resonator filter – HFSS.

5. Practical implementation of the three-band microstrip BPF

The practical implementation of a three-band microstrip bandpass filter involves turning the theoretical design into a functional device. This process includes detailed design, selecting appropriate materials, and manufacturing components on the dielectric substrate. The ultimate goal is to create a filter that meets performance specifications and is ready for integration into the final system or device.

This filter represents a complex and sophisticated device developed to meet specific requirements of the RF or microwave application in which it is used. To physically realize the filter, the PCB manufacturing line at the NUMELEC Research Center was used, and the model was imported into LPKF CircuitPro software as shown in Figure 8, where each component is assigned to its specific layer. In Figure 9, the physical implementation of a three-band microstrip resonator filter can be observed, constructed on a high-quality substrate known as RT/Duroid 5880.[8]

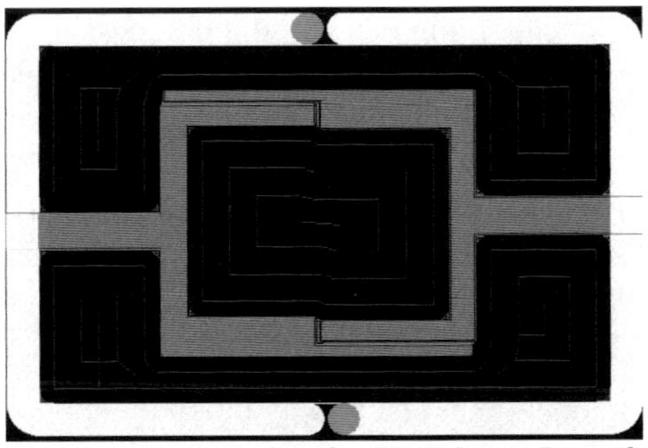

Figure 8. The geometry of the filter implemented in the LPKF CircuitPro software of the S103 plotter.

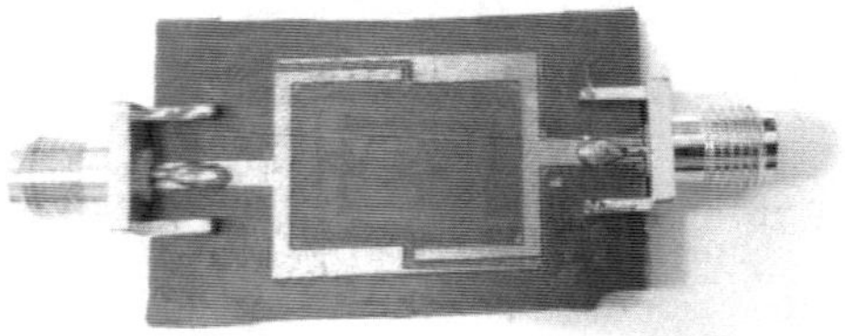

Figure 9. The microstrip resonator filter with three frequency bands physically constructed on RT/Duroid 5880.

Measuring the S parameters using a vector network analyzer (VNA) like the ZNL20 for a three-band bandpass filter (BPF) is an important process in characterizing and evaluating the performance of this type of filter, as shown in Figure 10. To conduct measurements accurately, the following aspects are taken into consideration:

Firstly, the ZNL20 vector network analyzer is configured, and appropriate connections are established between the VNA and the three-band BPF. The filter is connected to the VNA using a set of high-quality cables and connectors to minimize signal loss and ensure a stable connection. The frequency range selected is from 1 GHz to 8 GHz with a 1 kHz step and 1000 measurement points. Before commencing actual measurements, the VNA is calibrated [15].

Calibration is performed using the ZN-Z135 calibration kit for all necessary modes (Open, Short, Through) for the cables used to correct measurement errors introduced by the cables and connectors. This step is essential for obtaining precise results.

The S parameters include S_{11} (reflection at port 1), S_{12} (transmission from port 1 to port 2), S_{21} (transmission from port 2 to port 1), and S_{22} (reflection at port 2). These parameters provide information about the filter's behavior at different frequencies.

The results obtained from measurements and simulations, both modeled and practical outcomes from the vector network analyzer (VNA), are in accordance. The modeled results represent the theoretical predictions of the filter's behavior obtained through simulations performed earlier in the simulation software, ANSYS HFSS. These serve as reference data and provide a basis for understanding how the filter should ideally function in theory [15].

On the other hand, the practical results obtained with the VNA reflect the actual behavior of the filter under laboratory conditions. These measurements are used to validate and verify the effective operation of the device under real operating conditions. The results indicate a

IOP Conf. Series: Materials Science and Engineering **1320** (2024) 012006 doi:10.1088/1757-899X/1320/1/012006

correspondence between the theoretical model and the practical implementation of the filter, as shown in Figure 11. This suggests that the filter largely meets the design requirements and that the simulations provided an accurate prediction of its behavior.

Any significant differences between the modeled and practical results could indicate the need for further adjustments or optimizations of the filter to ensure that it fully meets the performance requirements and operates effectively at frequencies of 2.4 GHz, 3.5 GHz, and 5.2 GHz.

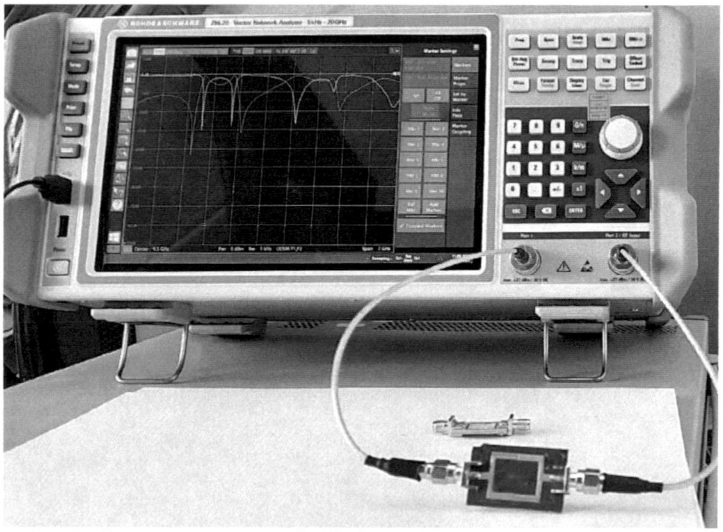

Figure 10. Measuring the S parameters using the VNA ZNL20.

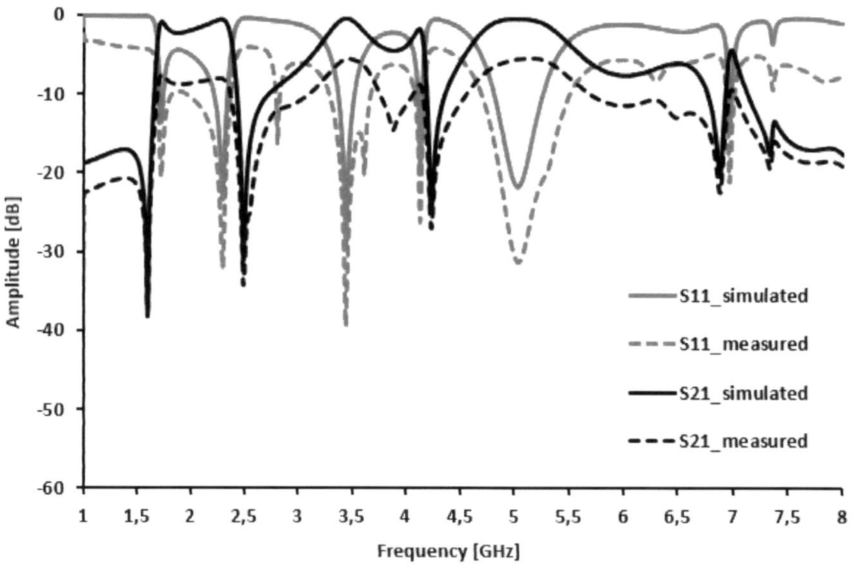

Figure 11. Comparing the results obtained from simulations and measurements of the designed three-band filter.

6. Conclusions

The microstrip three-band frequency bandpass filter, composed of two resonators, offers the advantage of being more compact and easier to fabricate compared to designs with multiple resonators.

By examining the variation of the quality factor and coupling coefficient based on dielectric thickness modification to optimize the performance of a microstrip bandpass filter at frequencies of 2.4 GHz, 3.5 GHz, and 5.8 GHz, it was found that the optimal thickness resulting from the analysis is 1.27 mm.

The designed filter operates at the 2.4 GHz band, commonly used in Wi-Fi networks and Bluetooth communications. The 3.5 GHz frequency is utilized in 5G networks for advanced mobile communication services, while the 5.2 GHz band is employed in high-performance Wi-Fi technologies and radar systems. The electric field is intense around the resonators, as these are the areas where the signal is amplified and directed to achieve the desired filtering characteristics.

The results obtained through modeling and experimentation are similar, indicating a proper validation of the model and simulation method used in this study. Table 3 provides a comparative analysis between the proposed filter and similar filters from the literature, considering their respective parameters.

Table 3. Comparison of the proposed three-band microstrip bandpass filters with other works from the literature.

Ref.	Frequency band [GHz]	L_{return} [dB]	$L_{insertion}$ [dB]	Dielectric ε_r/h[mm]	Dimension [mm^2]
[9]	2.5|3.5|5.8	32|31|32	0.8|2.3|2.4	RT 5880:2.2/0.78	30.7x12.3
[10]	0.9|2.4|3.5	11|21|18	0.85|1.1|1.1	Duroid:10.5/1.27	11.5x12.3
[11]	1.8|4.6|7.6	35|10|17	0.28|0.8|1.9	Epo_glass:4.4/1.57	22.6x12.9
[12]	2.4|3.5|5.2	15|13|20	1.5|2.54|1.16	RT 5880:2.2/1.57	21.1x9.12
[13]	2.4|3.5|5.2	18|16|13	2.0|2.4|1.7	2.55/0.8	24x22
Prop.	2.4|3.5|5.2	20|15|26	0.51|1.09|1	RT 5880:2.2/1.57	19.1x14.2

Upon analyzing the results, it is evident that the resulting size of the proposed filter is the smallest, while maintaining similar dielectric properties, with both return losses and insertion losses exhibiting among the best values.

References

[1] Jamil, S. H. S., Karim, S. S. A., and Murad, N. A. (2020). "Design and Simulation of Dual-Band Microstrip Filter Using Coupled Resonators," International Journal of Microwave and Wireless Technologies, 12(6), pp. 522-528.

[2] Bhat, Z. A., Sheikh, J. A., Khan, S. D., Rehman, R., and Ashraf, S. (2021). "Compact and Novel Coupled Line Microstrip Bandpass Filter Based on Stepped Impedance Resonators for Millimetre-Wave Communications," Frequenz -Berlin-, 75(5-6), February.

[3] Pawase, C. J., and Chang, K. (2023). "5G-NR Physical Layer-Based Solutions to Support High Mobility in 6G Non-Terrestrial Networks," Drones, 7, 176.

[4] Pacurar, C., Topa, V., Giurgiuman, A., Munteanu, C., Constantinescu, C., Gliga, M., and Andreica, S. (2021). "High Frequency Analysis and Optimization of Planar Spiral Inductors Used in Microelectronic Circuits," Electronics Journal, 10(23), 2897. ISSN: 2079-9292, IF: 2.397, November 23.

[5] Andreica, S., Munteanu, C., Gliga, M., Pacurar, C., Giurgiuman, A., and Constantinescu, C. (2023). "Development and Optimization of a Broadside-Coupled Dual-Band Microstrip Bandpass Filter for Wireless Communication Systems," 2023 10th International Conference on Modern Power Systems (MPS), DOI: 10.1109/MPS58874.2023.10187504, ISBN: 979-8-3503-2682-6, June 21-23.

[6] Constantinescu, C., Pacurar, C., Giurgiuman, A., Munteanu, C., Andreica, S., and Gliga, M. (2023). "High Gain Improved Planar Yagi Uda Antenna for 2.4 GHz Applications and Its Influence on Human Tissues," Microwave Antennas: Theory, Material, Technology, Design and Application, Appl. Sci., 13(11), 6678.

[7] Li, H., Li, Y., and Li, Y. (2019). "Design of a Dual-Band Microstrip Filter with High Selectivity and Stopband Rejection," IET Microwaves, Antennas & Propagation, 13(7), pp. 1035-1040.

[8] Andreica, S., Pacurar, C., Ţopa, V., Racasan, A., Constantinescu, C., and Gliga, M. (2017). "The Analysis of the Multilayer Spiral Inductors Parameters at High Frequency," International Conference on Modern Power Systems (MPS), Cluj-Napoca, Romania, ISBN 978-1-5090-6565-3/17, DOI: 10.1109/MPS.2017.7974429, June 6-9.

[9] Wei, F., Guo, Y. Jay, Qin, P.-Y., and Shi, X. Wei. (2015). "Compact Balanced Dual- and Tri-band Bandpass Filters Based on Stub Loaded Resonators," IEEE Microwave and Wireless Components Letters, 25(2), pp. 76-78, February

[10] A. M. Elelimy, A. M. El-Tager, A. G. Sobih and M. H. Abdel-Azeem, "Novel compact dual-mode tri-band bandpass filter for WiMAX & GSM applications," 2012 7th European Microwave Integrated Circuit Conference, Amsterdam, Netherlands, 2012, pp. 739-742.

[11] D. Rabahallah, M. Challal and N. Talaharis, "Tri-band microstrip bandpass filters for GSM and WiMAX applications," 2015 4th International Conference on Electrical Engineering (ICEE), Boumerdes, Algeria, 2015, pp. 1-4.

[12] N. Janković, R. Geschke and V. Crnojević-Bengin, "Compact Tri-Band Bandpass and Bandstop Filters Based on Hilbert-Fork Resonators," Microwave and Wireless Components Letters, volume 23, no.6, pp. 282-284, June 2013.

[13] F. -C. Chen and Q. -X. Chu, "Design of Compact Tri-Band Bandpass Filters Using Assembled Resonators," in IEEE Transactions on Microwave Theory and Techniques, vol. 57, no. 1, pp. 165-171, Jan. 2009.

[14] C. Pacurar, V. Topa, A. Giurgiuman, C. Munteanu, C. Constantinescu, M. Gliga, S. Andreica, The influence of the patch antennas emissions on the human head, IOP Conference Series: Materials Science and Engineering,ICEMS-BIOMED-2022, DOI: 10.1088/1757-899X/1254/1/012018, 2022.

[15] C. Constantinescu , C. Munteanu , C. Pacurar, A. Racasan, M. Gliga, S. Andreica, High Frequency Analysis of Bowtie Antennas, 2019 11th International Symposium on Advanced Topics in Electrical Engineering, ATEE 2019, Bucharest, Romania, 28-30 March 2019, ISBN: 978-147997514-3, DOI 10.1109/ATEE.2019.8724972, WOS: 000475904500129, 2019.

Evaluation of the head exposure to a static magnetic field while walking around 1.5T and 7T MRI magnets using single and spatially distributed Hall probes

Patryk Zradziński[1], Jolanta Karpowicz[1*], Thomas Quirin[2], Dominic Jeker[2], Joris Pascal[2], Andrzej Stępniewski[3] and Krzysztof Gryz[1]

[1] Department of Physical Hazards, Central Institute for Labour Protection – National Research Institute (CIOP-PIB), Warszawa, Poland
[2] Institute for Medical Engineering and Medical Informatics, School of Life Sciences, University of Applied Sciences and Arts Northwestern Switzerland (FHNW), Muttenz, Switzerland
[3] Center ECOTECH-COMPLEX, Maria Curie-Sklodowska University, Lublin, Poland

*E-mail: jokar@ciop.pl

Abstract. The aim of this study was to compare the use of static magnetic field (SMF) body-worn exposure monitors equipped with single or spatially distributed probes. The SMF exposure monitor allow to study the time-pattern of head movement-related exposure to SMF (B, mT; dB/dt, mT/s) while active near 1.5 T and 7 T magnetic resonance imaging (MRI) magnets. When analysing the distribution of B-field samples recorded during a simple walk around MRI magnets, using relevant statistical tests it was found that the results recorded by various probes are statistically significantly different. The observed scale of differences between the results from various probes revealed the possible misclassification of localised exposure of particular head sections when only a single probe is used to evaluate exposure (especially with respect to dB/dt values). Larger differences between probes were found in recordings near MRI magnets of smaller dimensions (1.5 T).

1. Introduction

Movement-related symptoms experienced by magnetic resonance imaging (MRI) workers exposed to a static magnetic field (SMF), such as vertigo or a loss of balance, need attention because of safety issues [1-6]. Better and more systematic knowledge is still needed in this field, including the relevant parameterisation of worker exposure during various real activities, or exposure to volunteers while performing research tasks.

The majority of studies already performed on exposure to MRI SMF evaluated exposure on the basis of measurements from a single body-worn probe [1-7]. The results of evaluating exposure in this way suffer from the spatial inhomogeneity of SMF near MRI magnets affecting the human body. This inhomogeneity may be more than 5-times increase of B-field over a volume of half a metre in diameter, i.e., over the volume of space comparable to the human body cross-section. To enhance MRI exposure monitoring studies, a solution may be to evaluate exposure on the basis of measurements from multiple points.

The aim of this study is therefore to compare exposure to SMF during activities close to MRI magnets, measured respectively with a body-worn single-probe exposure monitor and with a monitor with spatially distributed probe.

2. Methods

Measurements were performed using a versatile exposure monitor that can be read as a single probe or as spatially distributed probe, covering up to 8 measurement points. These were used in evaluating the time-pattern of movement-related exposure to SMF experienced by the volunteer's head. Test measurement sessions were performed during a simple walk near MRI magnets. During these sessions, the volunteer walked around the magnet, moving near to the magnet housing with stops at the 8 corners of the housing (A-H positions: near the patient's table, in the front and the back, on the right and left side; as well as near the side magnet corners, in the front and back, on the right and left side, as can be seen at figure 1).

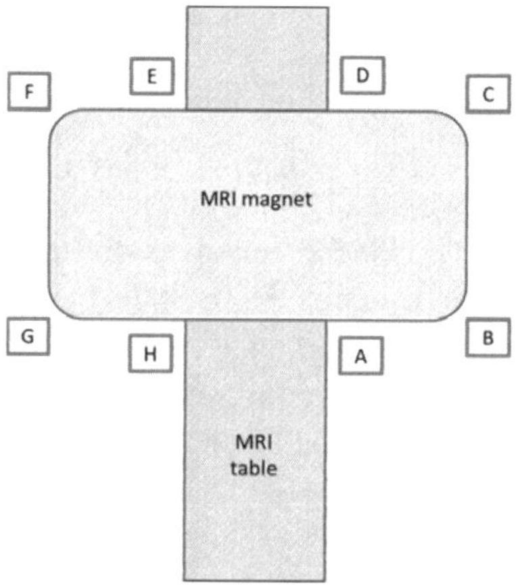

Figure 1. The scheme of the positions (A-H) of SMF exposure monitoring near MRI magnet.

The SMF exposure monitor used in this study is based on Hall-effect sensors [8]. Hall-effect sensors are currently the most suitable magnetometer type for SMF measurements, although emerging quantum sensors [9] could change this in the coming years. Hall sensors allow SMF dosimetry measurements mainly thanks to their high dynamic range, up to several T. They also take advantage of their compatibility with CMOS technology, which makes them small, low cost, and easily employing with standard electronics. The exposure monitor consists of a flexible PCB instrumented with Hall-effect sensors mounted on safety glasses [10]. Eight Hall sensors (S1-S8), three dimensional (i.e. using three directional orthogonally positioned independent Hall components, X, Y, Z), are located around the head: two sensors near the right side of the head (S1, S2), four sensors in front of the eyes (S3-S6) and two sensors near the left side of the head (S7, S8). The used 3D Hall sensors are TMAG5273A2 (Texas Instrument, USA) which feature a dynamic range of ± 266 mT on each axis for a sensitivity of 125 Least Significant Bit (LSB) per mT. The eight sensors are synchronously recorded with a 3 Hz sampling rate, i.e. fast enough to follow changes in the amplitude and orientation of the magnetic vector resulting from SMF and affecting the user of the exposure monitor (monitored worker or volunteer participating in research task) who moves normally. For more details, see SMF goggles-monitor technical characteristics published in [10].

Usually the module of B-field vector (BM) is used to characterise human exposure, derived from recorded values of orthogonal XYZ components. The dynamic changes in B-field exposure over time (dB/dt, in mT/s) is also of interest when human safety is considered near the MRI magnets, where movement-related component of SMF exposure effects in the human body may be significant [1-6,10]. After a measurement session where all data from the exposure monitor are collected, the distribution of B-field exposure (results from the measurements of six 3D Hall sensors evenly located around the head: two sensors near the right side of the head (S1, S2), two sensors in front of the eyes (S4, S6) and two sensors near the left side of the head (S7, S8)) is compared to single point measurements (based on recordings from the sensor located at the right side of the head – the S2 sensor). The distribution of the results obtained in a particular measurement session is characterised by the median value (Med), the 5-95 percentile range, and the minimum – maximum value range (MMR).

The B-field exposure samples were recorded using the goggles-monitor worn by a volunteer walking around the MRI magnets with a main field level of 1.5 T and 7 T. The 7 T device is equipped with a significantly larger magnet, which drastically reduces the level of the SMF accessible to humans during its regular use in the MRI unit, compared to the main field level inside the magnet, where the patient is positioned during an examination. To characterise the differences between the single point and spatially distributed measurements, the following scenarios were used:

(1) the volunteer's walk around the 7 T magnet, moving near to the magnet housing, with stops at its 8 corners (near the patient's table, in the front and back, to the right and left side; as well as near the corners of the side magnets, in the front and back, to the right and left side: A-H positions following the scheme at figure 1)
(2) the volunteer's walk around the 1.5 T magnet, moving near to the magnet housing only along its right side, with stops at its 4 mentioned corners (near the patient's table, in the front and back, on the right side; as well as near the corners of the side magnets, in the front and back, on the right side: A-D positions following the scheme at figure 1).

3. Results

An example of the time pattern of B-field recorded near the 1.5 T MRI magnet (time pattern of a vertical component (BZ) and of a module (BM) of the head exposure) is shown in figure 2.

The distribution of SMF samples recorded during the walk in particular measurement sessions is characterised in figures 3-5, where: (i) the distribution of the SMF values recorded by a particular sensor characterise the distribution of the head exposure on each measurement session, when particular sensor is monitoring this exposure, and (ii) the distribution of the relative values of exposure to SMF recorded by a particular sensor, normalised to the values recorded by the reference S2 sensor (S1vS2, S4vS2, ...), characterise the spatial distribution of the head exposure along the instantaneous recording time points, when particular pair of sensors are monitoring this exposure.

(a)

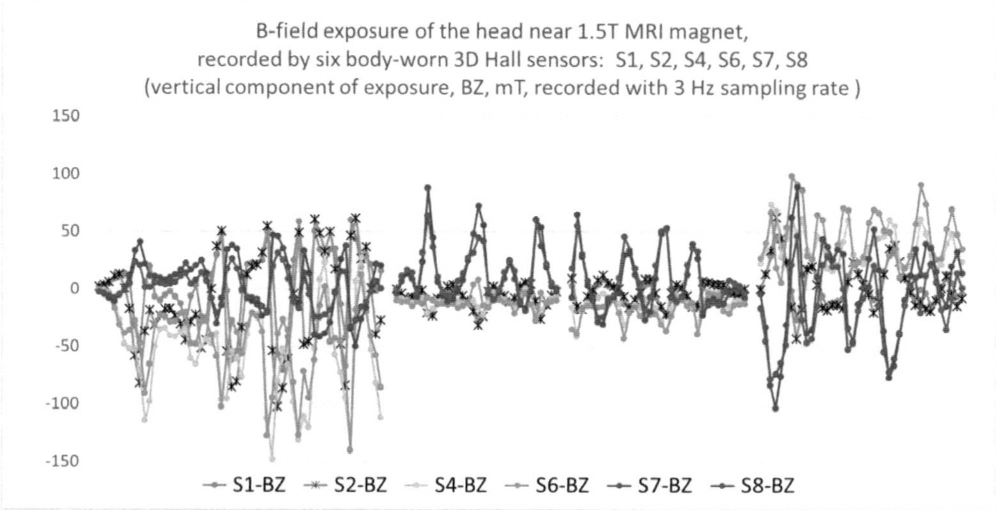

(b)

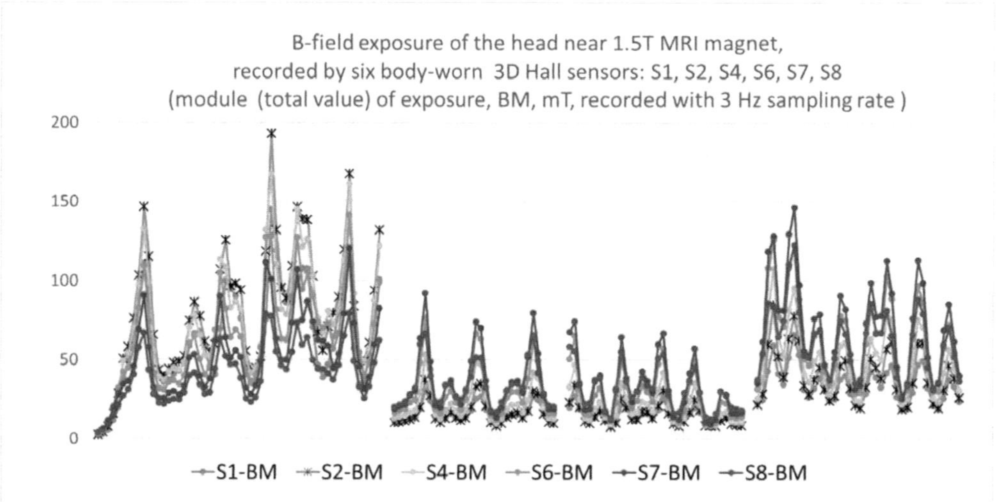

Figure 2. The example of B-field exposure recorded by the spatially distributed Hall probes (composed from six 3D sensors: S1, S2, S4, S6, S7 and S8) while walking near a 1.5 T MRI magnet, where the results of measurements by the referenced single Hall probe (S2) is marked by black asterisks: (a) time pattern of a vertical component (BZ) of the head exposure to SMF; (b) time pattern of a module (BM) of the head exposure to SMF.

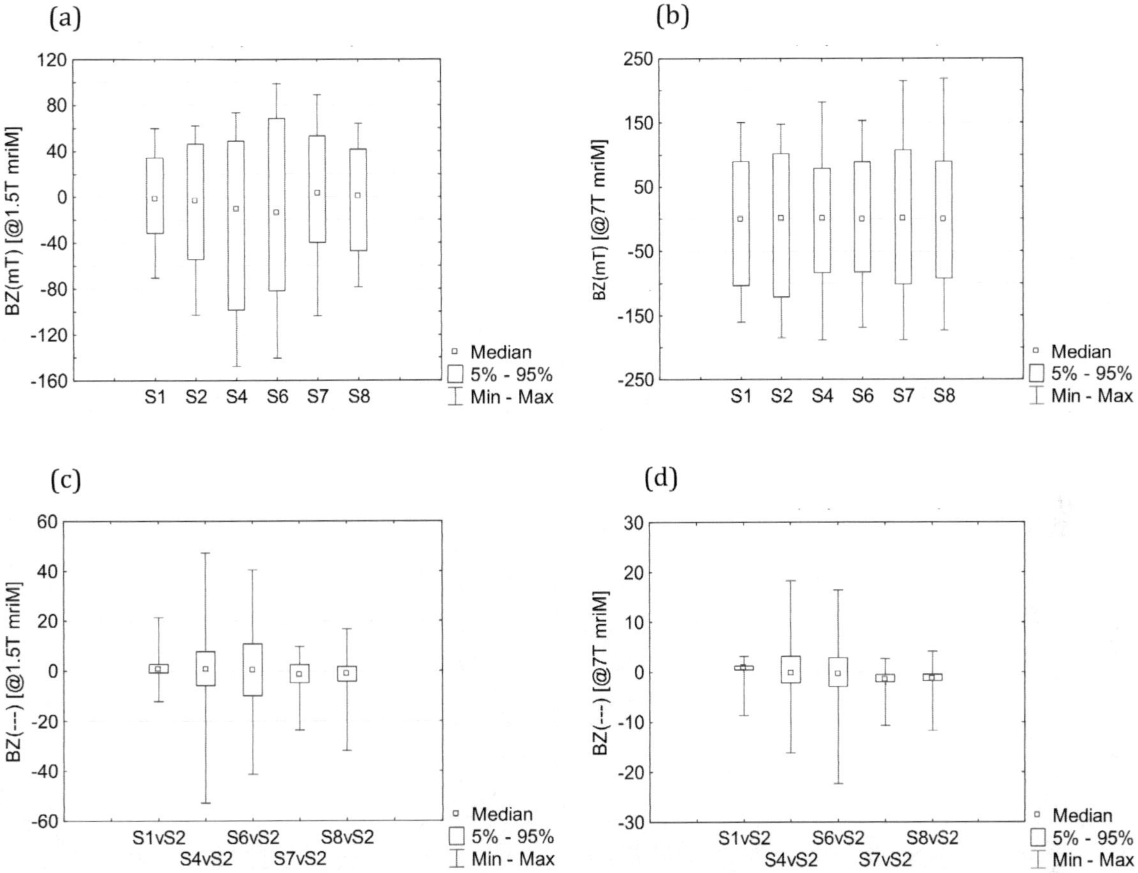

Figure 3. The distribution of SMF samples of B-field exposure of the head (vertical component, BZ value of exposure, recorded by particular component of the spatially distributed Hall probe: S1, S2, S4, S6, S7 and S8) while walking near a 1.5 T MRI magnet and 7 T MRI magnet; the reference single Hall probe is marked as S2: (a, b) the distribution of the SMF values recorded by a particular sensor (BM); (c, d) the distribution of the relative values of exposure to SMF recorded by a particular sensor, normalised to the values recorded by the reference S2 sensor (S1vS2, S4vS2, ...); (dot – median value; box – 5-95 percentile range; whiskers – MMR; mriM – MRI magnet).

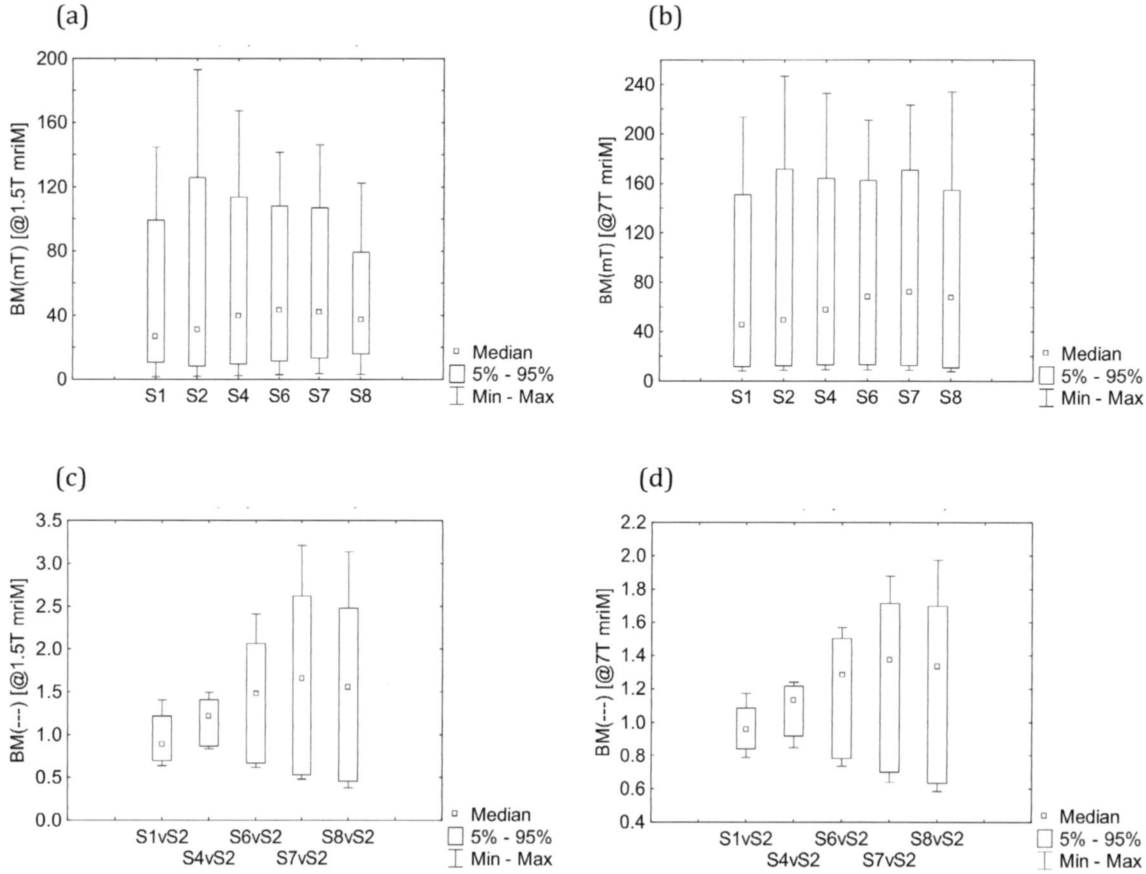

Figure 4. The distribution of SMF samples of B-field exposure of the head (the module, BM value of exposure recorded by a particular component of the spatially distributed Hall probe: S1, S2, S4, S6, S7 and S8) while walking near a 1.5 T MRI magnet and 7 T MRI magnet; the reference single Hall probe is marked as S2: (a, b) the distribution of the SMF values recorded by a particular sensor (BM); (c, d) the distribution of the relative values of exposure to SMF recorded by a particular sensor, normalised to the values recorded by the reference S2 sensor (S1vS2, S4vS2, ...); (dot – median value; box – 5-95 percentile range; whiskers – MMR; mriM – MRI magnet).

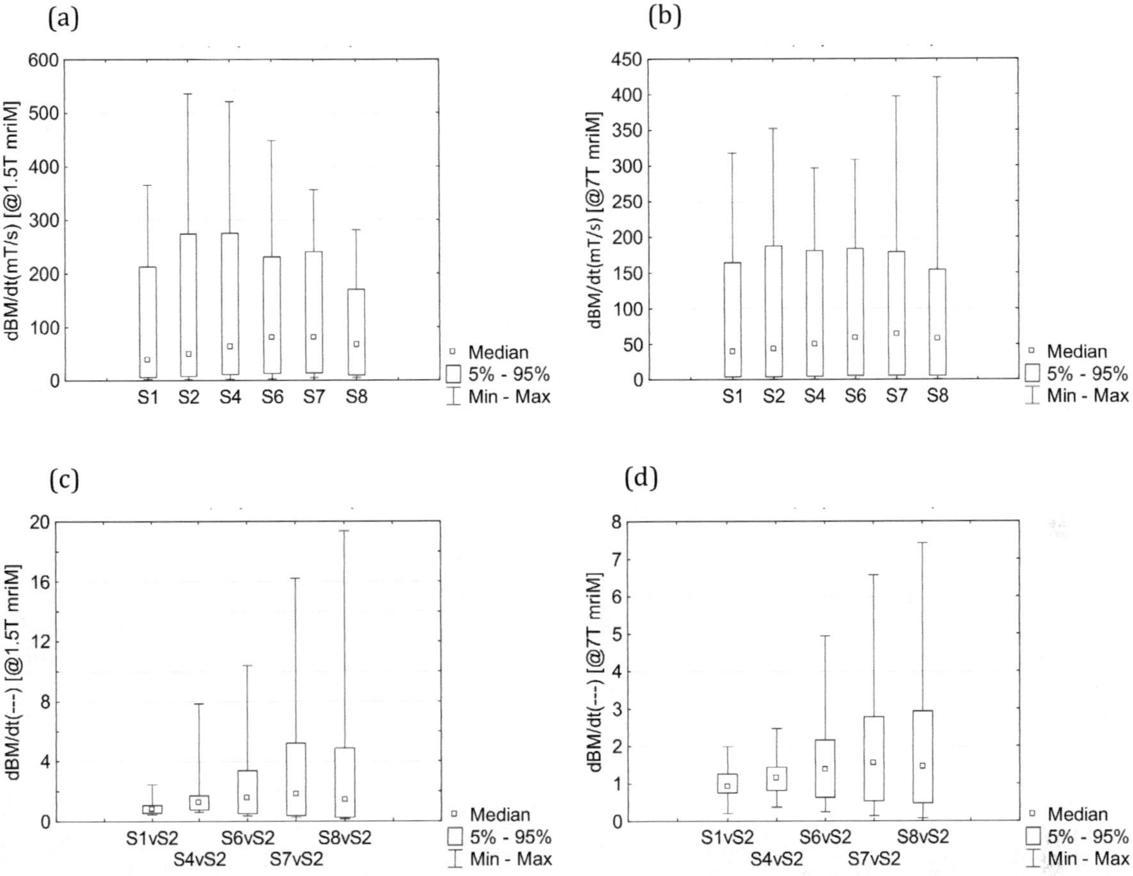

Figure 5. The distribution of SMF samples of B-field exposure of the head (the module, BM value of exposure recorded by a particular component of the spatially distributed Hall probe: S1, S2, S4, S6, S7 and S8) while walking near a 1.5 T MRI magnet and 7 T MRI magnet; the reference single Hall probe is marked as S2: (a, b) the distribution of the dynamic movement-related exposure pattern of the head exposure to SMF recorded by a particular sensor (dBM/dt); (c, d) the distribution of the relative dynamic movement-related exposure pattern of the head exposure to SMF recorded by a particular sensor, normalised to the values recorded by the reference S2 sensor (S1vS2, S4vS2, ...); (dot – median value; box – 5-95 percentile range; whiskers – MMR; mriM – MRI magnet).

4. Discussion and conclusions

The main differences between the analysed SMF exposure scenarios include: (i) the 7 T magnet is larger, meaning that the SMF accessible near its housing is more homogeneous than near the 1.5 T magnet; and (ii) activities during measurements near the 7 T magnet covered all combinations of SMF polarisation in various locations around the magnet (scenario of exposure to SMF of symmetric distribution of its polarisation during the measurement session), when activities during measurements near the 1.5 T magnet covered only one side of the magnet housing (scenario of exposure to SMF of asymmetric distribution of its polarisation during the measurement session – such distribution of exposure is very common during real activities of MRI workers, moving between the MRI table and the storage shelf containing MRI accessories, which is located at right or left side of the magnet).

Analysing the distribution of B-field exposure samples recorded using the goggles-mounted exposure monitor worn by a volunteer walking around the MRI magnet, significant differences were found between the single and spatially distributed measurement results. Nearly 10-times differences in the 5-95 percentile ranges were observed in the distributions of relative exposure patterns from particular pairs of elements of the distributed Hall probe (figure 3-5) – for example: figure 3c, S1vS2 compared with S6vS2, and figure 4c, S1vS2 compared with S7vS2. The maximum ratio between the B-field module values (BM) recorded by individual components and the reference single probe has values exceeding 3 (figure 4c: S7vS2 and S8vS2), where several decades of differences were observed in case of the single polarisation sensor (figure 3c: S4vS2 and S6vS2) and in the movement-related exposure pattern (figure 5c: S7vS2 and S8vS2).

Analysing the distribution of B-field samples recorded during the simple walk around MRI magnets, using relevant statistical tests (non-parametric ANOVA Kruskal–Wallis test for independent groups with multiple comparison of mean ranks; $p = 0.05$), it was found that the results recorded by various elements of a distributed probe (S1-S8) are statistically significantly different. Larger differences between probes were found in recordings near the MRI magnet of smaller dimensions (1.5T: figure 3a, 4a and 5a) than near the larger magnet (7T: figure 3b, 4b and 5b). The exposure scenario of asymmetric polarisation distribution (near the 1.5 T magnet) also caused larger differences in median and 5-95 percentile range values between particular sensors (S1-S8).

Significant differences in B values recorded by both type of measurements, in dBM/dt values (movement-related module of exposure), as well as in BZ values (single polarisation components of exposure), may reach high values in several per cent of samples (above 95th percentile – figure 3c,d, 5c,d), suggesting the possibility of different classification of localised head exposure when different measurement probes are used. This may also have an impact on the procedure of analysing of the human exposure pattern, because the electrodynamic effects of movement-related exposure near MRI magnets are sensitive to the polarisation of B-field that affects the body.

The observed pattern of exposure in the real MRI environment strongly supports the use of distributed measurements in further studies on movement-related symptoms, which may improve the safety of MRI personnel as a better understanding of the nature of exposure cases producing such symptoms can be worked out. The spatially distributed Hall probe (for example a set of Hall sensors supported by goggles) may improve the monitoring of the SMF exposure evaluation. Such sensors may be located close to the critical areas for the human body (such as the eyes, temples, etc.,) and also make it possible to calculate spatial gradients of SMF exposure across the head, which is not assessable from the results of single point monitoring of SMF exposure.

Acknowledgements: Results of the "Governmental Programme for Improvement of Safety and Working Conditions", funded by state services of the Ministry of Family, Labour and Social Policy (task no. 3.ZS.14, EM-Centrum)". The CIOP-PIB is the Programme's main co-ordinator.

References

[1] de Vocht F, van Drooge H, Engels H, Kromhout H. 2006. Exposure, health complaints and cognitive performance among employees of an MRI scanners manufacturing department. J Magn Reson Imaging, 23(2):197-204.
[2] de Vocht F, Stevens T, Glover P, Sunderland A, Gowland P, Kromhout H. 2007. Cognitive effects of head-movements in stray fields generated by a 7 Tesla whole-body MRI magnet. Bioelectromagnetics, 28(4):247-255.
[3] Scientific Committee on Emerging and Newly Identified Health Risks (SCENIHR). Opinion on Potential health effects of exposure to electromagnetic fields (EMF). European Commission: Luxembourg, 2015. Available online: http://ec.europa.eu/health/scientific_committees/emerging/docs/scenihr_o_041.pdf. (accessed on 6 December 2021).
[4] Bongers S, Slottje P, Kromhout H. 2018. Long-term exposure to static magnetic fields in MRI manufacturing and risk of developing hypertension, OEM; 73 (Suppl 1): A213-A214. https://doi.org/10.1136/oemed-2018-103951.592.
[5] Fatahi M. *et al.* 2017. Evaluation of exposure to (ultra) high static magnetic fields during activities around human MRI scanners. Magn. Reson. Mater. Phys. Biol. Med., 30, 255–264.
[6] Delmas A, Weber N, Piffre J, Pasquier C, Felblinger J, and Vuissoz P-A. 2017. MRI 'exposimetry': how to analyze, compare and represent worker exposure to static magnetic field? Radiat. Prot. Dosimetry, 177(4):415–423, doi: 10.1093/rpd/ncx060.
[7] Sklinda K, Karpowicz J, Stępniewski A. 2022. Electromagnetic Exposure of Personnel Involved in Cardiac MRI Examinations in 1.5T, 3T and 7T Scanners. IJERPH , 19, 76. https://doi.org/10.3390/ijerph19010076.
[8] Popovic RS. 2004. Hall Effect Devices, Second Edition. Institute of Physics.
[9] "Q.Ant Magnetometer", Available online: https://qant.com/magnetometer/ (accessed on 23 May 2024).
[10] Jeker D, Quirin T, Pascal J. 2023. Monitoring the exposure to magnetic fields of MRI workers using goggles integrating magnetometers, 2023 IEEE International Symposium on Medical Measurements and Applications (MeMeA), pp. 1–5. http://doi.org/10.1109/MeMeA57477.2023.10171916.

IOP Conf. Series: Materials Science and Engineering **1320** (2024) 012008 doi:10.1088/1757-899X/1320/1/012008

Multimodal evaluations of electromagnetic hazards while using MRI systems

Jolanta Karpowicz

Department of Physical Hazards, Central Institute for Labour Protection – National Research Institute (CIOP-PIB), Warszawa, Poland

*E-mail: jokar@ciop.pl

Abstract. At the heart of evaluating electromagnetic hazards near magnetic resonance imaging (MRI) systems used in medical diagnostics is to reveal the range of space surrounding the system that is affected by static magnetic field (SMF) to an extent potentially causing: (a) movement-related disturbances in human balance, which may lead to direct safety hazards for affected workers or patients; (b) a projectile effect in heavy ferromagnetic objects, which may create lethal hazards for workers and patients, as well as serious damage to the MRI scanner; (c) a projectile effect in small ferromagnetic objects, which may cause non-lethal hazards for workers or patients, for example by cutting the skin of a person hit by an object, though this is usually not dangerous to the MRI equipment. Additionally, thermal and electrodynamic effects in humans and equipment affected by time-varying MRI emissions (gradient and radiofrequency fields) need evaluation and prevention. The characteristics of static and time-varying electromagnetic fields near MRI systems and a multimodal evaluation of related hazards are discussed based on results of long-term environmental studies – using an exposure-oriented evaluation applying exposure limits or action levels, as well as a space-oriented evaluation applying the concept of Functional Intervention Levels (FIL). The newly tested applicability of autonomous EMF data loggers when evaluating these electromagnetic hazards near MRI systems is also considered.

1. Introduction

Magnetic resonance imaging (MRI) medical diagnostic systems involve the use of specific sources of electromagnetic field (EMF), namely: the strong or ultra-strong static magnetic field (Smf), the intermediate frequency electromagnetic field (known as the gradient field (GRemf)) and the radiofrequency electromagnetic field (RFemf), which cause the intentional patient's exposure during the acquisition of body scans using an MRI scanner (mriS) [2, 11, 16]. The level of Smf (homogeneous inside mriS) applied to patients (parameterised by magnetic flux density – B, expressed in Tesla – T) is the key parameter for the diagnostic power of an MRI system. In order to obtain the imaging (cross-sections) of the patient's body morphology, all these types of EMF are emitted by an MRI, and also affect the environment surrounding the MRI magnet (mriM), located inside the electromagnetically shielded MRI diagnostic room (mriDR), (figure 1). The organisation of work and the workplace inside the mriDR need to remember and respect the fact that, during the two-second walk to cover the two or three metres approaching the mriM, an adult's Smf exposure may rise 1000-times comparing to the levels at the mriDR entrance (or fall respectively), which also drives strong movement-related exposure to magnetic field varying in time [16-19, 26].

2. EMF exposure while using MRI systems

Because the superconducting and permanent magnet is active permanently (24 hours per day, 7 day per week, 52 weeks per year – as long as it is active), Smf affects everything and everyone present in proximity to the mriM. Because of various physical activities near an Smf source, no living organism is ever exposed to a homogeneous and constant in time Smf there. This means that, apart from the specific effects from Smf exposure observed near an mriM in any object, when that object changes its position there, additional exposure effects are observed, similar to those caused by time-varying EMF influence [2, 8, 26]. Because of that, when evaluating EMF hazards, a fourth type of exposure (EMF source) may also be considered – movement-related exposure to the magnetic field (MRmf).

Contrary to Smf (with the most typical 1.5 T or 3 T level of the main field, B0, affecting the patient inside the mriS), the time-varying GRemf (with acoustic frequencies ranging approximately 1-10 kHz) and RFemf (with the most typical frequencies at 63 MHz or 127 MHz) are emitted only while actually acquiring MRI scans. While scanning a patient, GRemf is emitted and accompanied by the secondary noise emission from the mriS. Usually, the patient is present inside the mriM at that time, but MRI personnel are usually controlling the functioning of a device from a computer console outside of the mriDR [11, 17, 19, 28]. In specific cases, such as with children or disabled patients, MRI personnel or others may also assist patients during the emission of time-varying EMF. Because of that, exposure to GRemf and RFemf must also be monitored and evaluated in the mriDR.

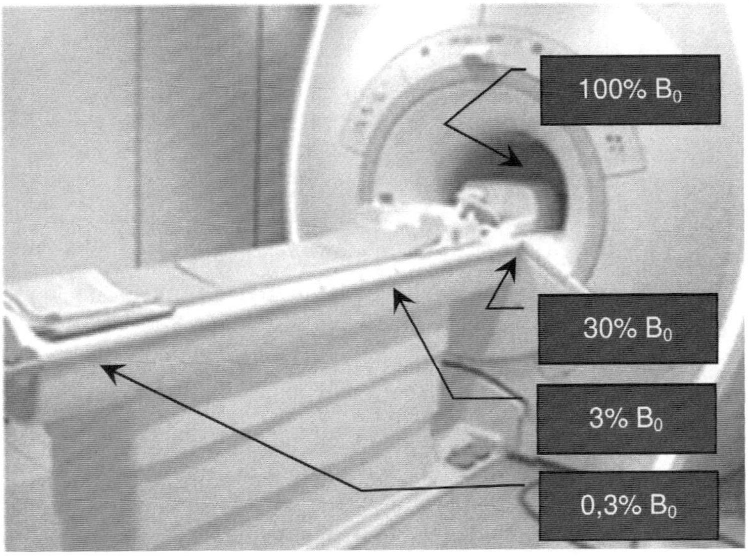

Figure 1. The most popular structure of an MRI scanner and the scheme of spatial distribution of the inhomogeneous static magnetic field inside an MRI diagnostic room (B0 – level of the main field inside the scanner; the most common today are scanners with 1.5T or 3T main field magnets).

IOP Conf. Series: Materials Science and Engineering **1320** (2024) 012008 doi:10.1088/1757-899X/1320/1/012008

While the protocol of MRI scans compilation is are dependent on the magnetic components of EMF used and affecting the patient (static, gradient and radiofrequency magnetic field), it must be remembered that the near-field EMF in the mriDR include both the electric and magnetic fields, and the evaluation must cover exposure to all these fields.

RFemf and GRemf are emitted in sequences of pulses, of dynamic and time-variations profiles which change significantly between various MRI systems and patient examination types, as well as during a single examination (figure 2) [2, 11, 15, 16].

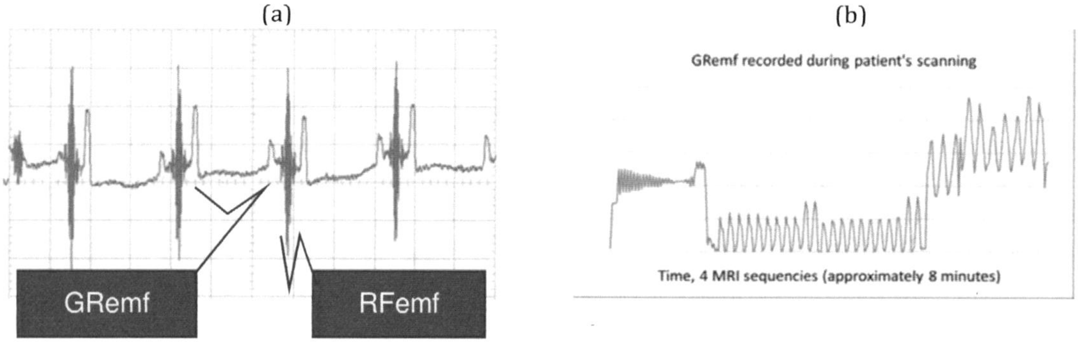

Figure 2. The profile of the time-variations of EMF emitted by an mriS, in the perspective of single RFemf and GRemf pulses and variations between diagnostic sequences during the patient's examination – an example from oscilloscopic monitoring (a) representing the shape and frequency of GRemf pulses harmonised with pulsed modulated RFemf, and the GRemf recording by low-frequency root-mean-square magnetic field data-logger (b), [linear non-calibrated dynamic vertical scale (a and b); horizontal scale: 5 ms/div (a) and 1.5 second sampling rate during 8 minutes (b)].

3. Electromagnetic hazards related to the use of MRI systems

3.1. Human exposure related hazards

Characterising various health and safety hazards in the environment in the mriDR is still a research priority in bioelectromagnetics and environmental engineering, as well as a core task in managing occupational hazards at work in medical facilities [2, 11, 17]. In view of the practical challenges the evaluation of these hazards entails, a multimodal approach is required due to the complex frequency and time pattern and the spatial distribution of EMF in the vicinity of the mriM, which also differ between the mriS from various manufacturers, the physical locations of the mriM inside the mriDR, and the purpose and method of use.

The study characterised in this paper has been worked out within the obligatory monitoring of exposure to EMF in the work environment, as required by labour law; in Poland, capturing frequency bands covering MRI RFemf since 1977, and GRemf and Smf since 1995 [16, 17]. Because of that, the significant development in MRI use in Poland over the last 30 years (from several low field mriS at the beginning, i.e. equipped with magnets of B0 ≤ 0.5T, up to several hundreds of devices of various diagnostic power, equipped with (0.1-7) T magnets operated nowadays), was assisted by studies focused on EMF exposure while using MRI systems, involving the Laboratory of Electromagnetic Hazards at CIOP-PIB since 1993 [8, 11, 16-21, 26].

Any object exposed to EMF near an mriM is affected by what are known as immediate exposure effects [2, 12-14, 24]:

- mechanical forces experienced by solid objects or liquids: a translation effect caused by heterogeneous Smf and proportional to B(dB/dx), or a rotation effect caused by homogeneous Smf and proportional to squared B,
- electric potentials induced in the object – an electrodynamic effect caused by Smf, GRemf, RFemf and MRmf,
- electric currents induced in the object by the induced electric potentials.

These EMF exposure effects may cause various physiological effects in the human body (depending on frequency, modulation, polarisation, level and duration of exposure) [2, 12-14, 24], including:

- electric stimulation in the nervous system (central and peripheral nervous systems, senses organs, muscles; even when there is no stimulation of receptors) – caused by Smf, GRemf and MRmf,
- thermal stress, which may be significant enough to cause thermal damage in tissues – caused by RFemf, but also strengthened by GRemf.

A variety of mechano- and electrodynamic effects caused by the influence of Smf (inhomogeneous or homogeneous) on objects and humans may lead to immediate significant health and safety hazards in the workplace, such as: mechanical damage from flying objects attracted to the magnet; significant disturbing effects in the sense of balance of any humans moving near the mriM, such as vertigo and loss of balance, which may disturb the work performance or even cause accidents, such as someone falling down [12-14]. The health effects of chronic (over the course of years) exposure to Smf are still under scientific consideration, taking into account the mentioned mechano- and electrodynamic effects of exposure, as well as the electromagnetic influence on chemical and biochemical processes in the body [3-11, 22-30]. This is because the strongest Smf exposure at present may be found only near the mriM, there are no supporting studies in that area from other workplaces.

The effects mentioned here are dependent on a variety of parameters characterising the static and dynamic influence of Smf, such as B-field level and spatial distribution (known as spatial gradients: dB/dx, dy, dz distribution), as well as dynamic changes of exposure caused by body movements dB/dt. All of these may be considered as characteristic of the module of B-field vector, or with respect to each of three-dimensional orthogonal components (xyz) of the B-field vector. Such a complex pattern of exposure, as shown at figure 3, for example, create huge challenges in the routine practice of occupational safety engineering.

3.2. Human exposure related hazards

Additionally, induced voltages, currents and heat may induce malfunctions in electronic devices of any type, for example patients' monitoring systems, medical body-worn or implanted devices (such as cardiac pacemakers, hearing stimulators or insulin delivery pumps) or any measurement devices, including EMF monitoring devices [12-14, 30]. The most common induced malfunctions near mriS include: damage to magnetic memory cards, the strong mechanical attraction of device to the mriM because of inner ferromagnetic components (for example batteries), the indication screen spontaneously switching off due to the impact of strong Smf or time-varying EMF, various typical malfunctions related to electromagnetic interferences (EMI), when the device is changes its position (e.g. in the case of body-worn devices) or when GRemf and RFemf are emitted by an mriS.

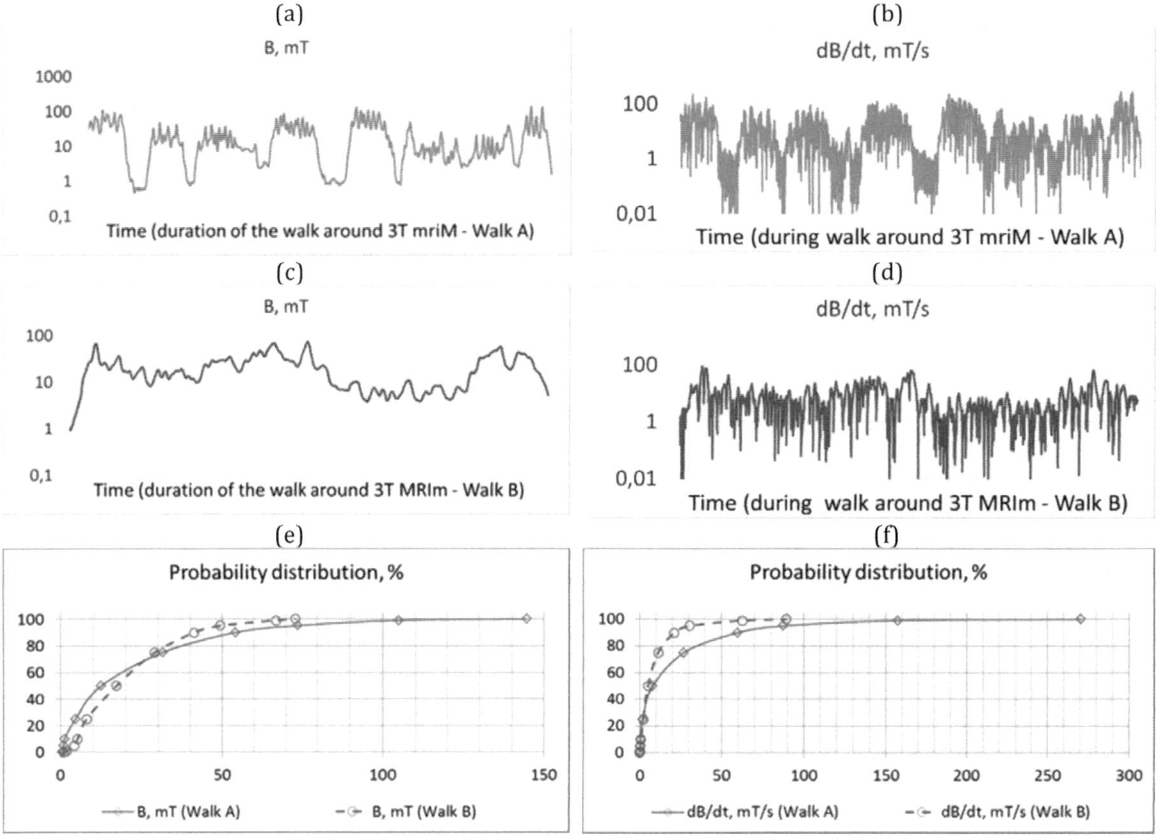

Figure 3. The example of the variability of Smf exposure profiles (a-d), when various movement strategies are applied near an mriM: walk A and walk B around the same 3T mriM, and the comparison of probability distribution of analysed parameters of the shown exposure profiles (e-f), (the head exposure monitored using a wearable Hall probe data logger).

The heat induced in a body-worn or implanted devices may cause also thermal damage to adjacent tissue [2, 30]. Because of these effects exerted on devices, any medical devices intended for regular use inside an mriDR have to be certified as being MRI-compatible (figure 4). Given that EMF monitors are usually designed for environmental measurements in regular work or publicly accessible places, the majority of them are not certified as MRI-compatible, or have warnings that they are not to be used in strong Smf (for example, because of the type of batteries used, which are not MRI-compatible).Various tests, performed during our MRI studies, revealed that in Smf of up to approximately 0.5 T, many types of autonomous electronic data-loggers developed for various kind of EMF studies may be successfully used inside an mriDR – with close attention to the safety issues related to the magneto-mechanical hazards near an mriM. For example, figure 2 showed an example of GRemf measurements using a data-logger developed for monitoring magnetic field exposure in the industrial environment. However, it must also be pointed out that measurements of pulsed GRemf and RFemf near MRI systems have some similarities to radar EMF emission measurements, which means that each measurement device needs the specific calibration of its dynamic output (because a regular calibration of indications of EMF monitors is performed in the sinusoidal continuous EMF of a particular frequency, while EMF evaluated in an mriDR is composed from three kinds of fields – CW Smf and pulsed GRemf, and pulse modulated RFemf – all affecting any measurement device simultaneously) [21]. Complex EMF exposure inside an mriDR may additionally cause false indications because of the possibly out-of-band sensitivity of the GRemf meter (i.e. an EMF meter designed for measurements covering the frequency band of GRemf) to RFemf exposure, as well as RFemf meter false indications caused by GRemf exposure.

	**MRI non-compatible object** (object hazardous near mriM / red label)
	**MRI compatible object** (object safe near mriM / green label)

Figure 4. The standardised labelling of MRI compatible and non-compatible objects.

3.3 Evaluation of electromagnetic hazards in mriDR

Several guidelines on Smf limits (regarding the levels of magnetic flux density, expressed in Tesla, and supported by limits on its time-derivative dB/dt, expressed in mT/s, or limits on weighted-peak value of MRmf, B, expressed in mT) have been published (and amended) over the past decades [12-14]. These are applicable in the process of evaluating various hazards at particular locations based on Smf measurements (in principle) [1, 11-20]. However, in practice, the use of such limits may be inconvenient because the spatial gradient of Smf near an mriM is very steep – with the level of a 10-fold increase as the distance to the magnet's housing decreases by just 50 cm, which is comparable to the space occupied by a worker's body performing any task in a "fixed" location.

The core evaluation of electromagnetic hazards near MRI systems used in medical diagnostics is to reveal the range of space in the surroundings affected by Smf to an extent that may potentially cause:

(a) movement-related disturbances in human balance, which may lead to direct safety hazards for affected workers or patients;

(b) a projectile effect in heavy ferromagnetic objects, which may cause lethal hazards for workers and patients, as well as serious damage to the mriS;

(c) a projectile effect in small ferromagnetic objects, which may cause non-lethal hazards for workers or patients, for example by cutting the skin of a person hit by a flying object, though this is usually not dangerous to the MRI equipment.

Precise parameterisation of these hazards needs a complex multimodal evaluation based on measurements covering various time- and spatial- characteristics of exposure experienced by workers nearby an mriM. However, based on the broad results of relevant environmental studies, the evaluation of the discussed hazards for practical use may be supplemented by Functional Intervention Levels (FIL), derived in the context of the spatial location of environments sensitive to particular hazards caused by Smf: FILa – for movement-related hazards, FILb – for projectile effects in heavy objects, and FILc – for projectile effects in small objects. By analysing the collected data, it was found that, with sufficient precision, the analysed hazards may be managed near 1.5T and 3T MRI scanners using the following description of FILs (figure 5):

- FILa – the space in front of/behind the MRI scanner, from its housing up to a distance of a quarter of the length of the MRI patient's table;
- FILb – the space in front of/behind the MRI scanner, from its housing up to a distance of half the length of the MRI patient's table and the space to the side of the magnet, between the magnet and walls of the diagnostic chamber;
- FILc – the space in front of/behind the MRI scanner, from its housing up to a distance of the entire length of the MRI patient's table and the space to the side of the magnet, between the magnet and the walls of the diagnostic chamber, or, to take a more practical approach, the entire interior of the diagnostic room.

The suggested practical approach (using the next modality in the evaluation of hazards in the context of space inside an mriDR) may help with managing serious safety hazards in the MRI environment and when providing safety education, especially for personnel frequently changing work tasks and locations (medical and non-medical), or entering MRI chambers only on unusual occasions (such as for the technical supervision of any installations inside).

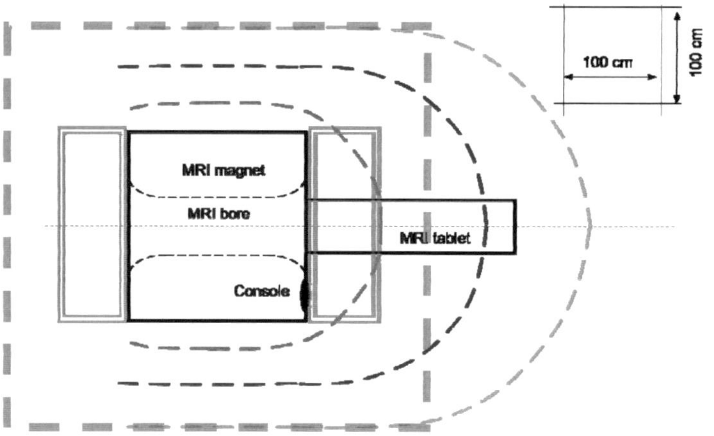

Figure 5. The scheme of typical spatial distribution of an inhomogeneous static magnetic field inside an MRI diagnostic room and the FILs concept; B-field isolines indicate the typical range near the mriM with exposure exceeding 50 mT (red line), 3 mT (blue line) and 0.5 mT (green line). This spatial distribution of typical Smf exposure in an mriDR supported the FILs distribution, drafted using bold bars: FILs – red bar with double-line border; FILb – blue bold bar with the dotted line border.

The suggested approach may also help with classifying the level of exposure of workers, as well as potential exposure effects caused by Smf, in epidemiological studies (for example, the nurses' tasks in many MRI diagnostic units are performed in exposure levels not exceeding the FILa level, whereas all radiographers need to be considered as exposed up to the level of the main field inside the MRI magnet's bore (i.e. up to 1.5T or 3T), where all the considered safety hazards may occur.

4. Conclusions

The characteristics of static and time-varying EMF exposure near MRI systems and the multimodal evaluation of related hazards experienced by workers nearby were discussed based on the results of extensive environmental studies.

The application of an exposure-oriented evaluation – using exposure limits or action levels, as well as a space-oriented evaluation – using the Functional Intervention Levels (FILs) concept were discussed. They were supplemented by considerations on the practical problems related to the relevant EMF measurements near MRI systems where specific strong and complex EMF influence on the measurement devices need to be taken into account. The newly tested applicability of autonomous EMF data loggers when evaluating these electromagnetic hazards near MRI systems was also considered.

A practical approach to evaluating and managing safety hazards in the MRI environment was developed using the "safety distances" approach, which is applied commonly in work-related safety guidelines. This classical approach was converted into the classification of environment inside mriDR based on the functional FILs derived in the context of the location of space at particular hazards: FILa – for movement-related hazards, FILb – for projectile effects in heavy objects and FILc – for projectile effects in small objects.

The application of the suggested easily understandable classification of the space near MRI system is mainly related to managing serious safety hazards in the MRI environment and safety education. It may also help classify exposure levels in epidemiological studies.

Abbreviations:
EMF - electromagnetic field
GRemf - gradient electromagnetic field
MRI - magnetic resonance imaging
MRmf - movement-related exposure to magnetic field
mriDR – MRI diagnostic room
mriM - MRI magnet
mriS - MRI scanner
RFemf - radiofrequency electromagnetic field
Smf - static magnetic field

Acknowledgments: The results of a research task (EM-Centrum: 3.ZS.14, 2.SP.10, 1.G.12) carried out within the scope of the sixth stage of the "Governmental Programme for Improvement of Safety and Working Conditions" supported in Poland within the scope of state services by the Ministry of Family, Labour and Social Policy (CIOP-PIB is the Programme's main co-ordinator). Research within the statutory activity of the CIOP-PIB (II-33), supported by the Ministry of Science and Higher Education, Poland (2014-2015).

Conflict of interest - the author declares that she has no conflict of interest.

References

[1] Andreuccetti D, Contessa GM, Falsaperla R, Lodato R., Pinto R, Zopetti N, Rossi P. (2013). Weighted-peak assessment of occupational exposure due to MRI gradient fields and movements in a nonhomogeneous static field. Med. Phys, 40

[2] Capstick M, McRobbie D, Hand J, Christ A, Kühn S, Hansson Mild K, et al. (2008). An investigation into occupational exposure to electromagnetic fields for personnel working with and around medical magnetic resonance imaging equipment. Report on Project VT/2007/017 of the European Commission, DG Employment, Social Affairs and Equal Opportunities. http://www.myesr.org/html/img/pool/VT2007017FinalReportv04.pdf last accessed 24 October, 2011.

[3] de Vocht F, van Wendel de Joode B, Engels H, Kromhout H. (2003). Neurobehavioral effects among subjects exposed to high static and gradient magnetic fields from a 1.5 Tesla magnetic resonance imaging system - a case-crossover pilot study. Magn Reson Med. 50(4):670-674.

[4] de Vocht F, van Drooge H, Engels H, Kromhout H. (2006). Exposure, health complaints and cognitive performance among employees of an MRI scanners manufacturing department. J Magn Reson Imaging, 23(2):197-204.

[5] de Vocht F, Stevens T, van Wendel de Joode B, Engels H, Kromhout H. (2006). Acute neurobehavioral effects of exposure to static magnetic fields: analyses of exposure-response relations. J Magn Reson Imaging, 23(3):291-297.

[6] de Vocht F, Stevens T, Glover P, Sunderland A, Gowland P, Kromhout H. (2007). Cognitive effects of head-movements in stray fields generated by a 7 Tesla whole-body MRI magnet. Bioelectromagnetics, 28(4):247-255.

[7] de Vocht F, Muller F, Engels H, Kromhout H. (2009). Personal exposure to static and time-varying magnetic fields during MRI system test procedures. JMRI 30:1223-1228.

[8] Fatahi M, Karpowicz J, Gryz K, Fattahi A, Rose G, Speck O. (2017). Evaluation of exposure to (ultra) high static magnetic fields during activities around human MRI scanners. Magnetic Resonance Materials in Physics, Biol. Med. (MAGMA) 30:255-264. doi: 10.1007/s10334-016-0602-z.

[9] Glover PM, Cavin I, Qian W, Bowtell R, Gowland PA. (2007). Magnetic-field-induced vertigo: a theoretical and experimental investigation. Bioelectromagnetics 28:349-361.

[10] Gowland and Glover (2014) Comment on ICNIRP Guidelines for Limiting Exposure to Electric Fields Induced by Movement of the Human Body in a Static Magnetic Field and by Time-varying Magnetic Fields Below 1 Hz; Health Physics, 107(3):261.

[11] Hansson Mild K, Hand J, Hietanen M, Gowland P, Karpowicz J, Keevil S, Lagroye I, van Rongen E, Scarfi MR, Wilén (2013). Exposure classification of MRI workers in epidemiological studies. Bioelectromagnetics 34(1):81–84.

[12] International Commission on Non-Ionising Radiation Protection (2009) Guidelines on limits of exposure to static magnetic fields. Health Phys 96:504-514

[13] International Commission on Non-Ionising Radiation Protection (2010) Guidelines for Limiting Exposure to Time-Varying Electric and Magnetic Fields (1 Hz - 100 kHz). Health Phys 99(6):818-836

[14] International Commission on Non-ionizing Radiation Protection (2014): Guidelines for limiting exposure to electric fields induced by movement of the human body in a static magnetic field and by time-varying magnetic field below 1 Hz. Health Physics 106(3), 418-425.

[15] Kanala S, Toivo T, Alanko T, Jokela K. (2009). Occupational exposure measurements of static and pulsed gradient fields in the vicinity of MRI scanners. Phys. Med. Biol., 54: 2243-57.

[16] Karpowicz J, Gryz K. (2006). Health Risk Assessment of Occupational Exposure to a Magnetic Field From Magnetic Resonance Imaging Devices. Int. Journal Occup. Safety and Ergon. (JOSE), vol. 12, No. 2, pp 155-167.

[17] Karpowicz J, Hietanen M, Gryz K. (2007). Occupational risk from static magnetic fields of MRI scanners. Environmentalist, vol 27(4), pp 533-538.

[18] Karpowicz J, Gryz K, Politański P, Zmyślony M. (2011). Exposure to static magnetic field and health hazards during operation of magnetic resonance scanners. Medycyna Pracy, 62(3): pp 309-321 (PL)

[19] Karpowicz J, Gryz K. (2013). The pattern of exposure to static magnetic field of nurses involved in activities related to contrast administration into patients diagnosed in 1.5 T MRI scanners. Electromagnetic Biology and Medicine, 32(2): 182–191.

[20] Karpowicz J, Gryz K. (2013). Experimental evaluation of ballistic hazards in imaging diagnostic center, Polish Journal of Radiology, 2013, 78(2): 31-37

[21] Karpowicz J, Bieńkowski P, Kieliszek J. (2016). Model of the minimum requirements regarding electric and magnetic field strength measurement devices for use in the near-field occupational exposure in compliance testing with respect to the requirements of European Directive 2013/35/EU," 2016 International Symposium on Electromagnetic Compatibility - EMC EUROPE, Wroclaw, Poland, pp. 668-671, doi: 10.1109/EMCEurope.2016.7739277.

[22] Lee JW, Kim MS, Kim YJ, Choi YJ, Lee Y, Chung HW. 2011. Genotoxic effects of 3 T magnetic resonance imaging in cultured human lymphocytes. Bioelectromagnetics. 32(7):535-542.

[23] Prato F, Thomas AW, Legros A, Robertson JA, Modolo J, Stodilka RZ, Demoor JM, Huda W (2012). MRI safety not scientifically proven. Science: 328, 568–569.

[24] SCENIHR [Scientific Committee on Emerging and Newly Identified Health Risks]. (2015). Opinion on potential health effects of exposure to electromagnetic fields (EMF). Luxembourg: European Commission. https://ec.europa.eu/health/scientific_committees/emerging/docs/scenihr_o_041.pdf.

[25] Simi S, Ballardin M, Casella M, De Marchi D, Hartwig V, Giovannetti G, Vanello N, Gabbriellini S, Landini L, Lombardi M. (2008). Is the genotoxic effect of magnetic resonance negligible? Low persistence of micronucleus frequency in lymphocytes of individuals after cardiac scan. Mutat. Res. 645(1-2):39-43.

[26] Sklinda K, Karpowicz J, Stępniewski A. (2022).Electromagnetic Exposure of Personnel Involved in Cardiac MRI Examinations in 1.5T, 3T and 7T Scanners. International Journal of Environmental Research and Public Health. 19(1):76. https://doi.org/10.3390/ijerph19010076

[27] Theysohn JM, Maderwald S, Kraff O, Moenninghoff C, Ladd ME, Ladd SC. 2008. Subjective acceptance of 7 Tesla MRI for human imaging. MAGMA 21(1-2):63-72.

[28] Wilén J, de Vocht JF (2011). Health complaints among nurses working near MRI scanners – a descriptive pilot study. Eur. J. Radiol. 80(2):510–513.

[29] Yamaguci-Sekino S, Nakai T, Imai S, Izawa S, Okuno T. (2014). Occupational exposure levels of static magnetic field during routine MRI examination in 3 T MR system. Bioelectromagnetic, 35: 70-75

[30] Zradziński P, Karpowicz J, Gryz K, Leszko W. (2018). Evaluation of the safety of users of active implantable medical devices (AIMD) in the working environment in terms of exposure to electromagnetic fields: practical approach to the requirements of European Directive 2013/35/EU. Int. J. Occup. Med. Environ. Health 31(6), 795–808.

IOP Conf. Series: Materials Science and Engineering **1320** (2024) 012009 doi:10.1088/1757-899X/1320/1/012009

Electromagnetic exposure of population in three cities of Romania at four years difference, based on data recorded by the fixed national radiofrequency monitoring network

D B Deaconescu[1], T Petrita[2], S Miclaus[3]

[1] Doctoral School of Electrical Engineering, Technical University of Cluj-Napoca, Cluj-Napoca, Romania

[2] Department of Communications, Politehnica University of Timisoara, Romania

[3] Department of Communications, IT & Cyb. Defense "Nicolae Bălcescu" Land Forces Academy Sibiu, Romania

deliadeaconescu@yahoo.com, teodor.petrita@gmail.com, simo.miclaus@gmail.com

Abstract. The blooming of wireless technologies and the deployment of the fifth generation (5G) of mobile communications have been accompanied by a significant increase in the risk perception related to human exposure to radio-frequency electromagnetic field (RF-EMF). Therefore, EMF monitoring has become necessary in areas with increased densification of the EMF sources. Present work made use of the raw data kindly provided by the National Authority for Management and Regulation in Communications (ANCOM) in Romania from the fixed EMF sensor network nationally deployed. We aimed to analyze the case of electric field level evolution after four years, in three cities. In each city a single monitoring station was chosen in the analysis. Database in the wide- and narrow-bandwidths covering the range (100 kHz – 7 GHz) was processed for whole years 2019 and 2023. Comparisons between locations and years were enabled by means of: a) descriptive statistics; b) time series analysis with Poincare plots for time-variability characterization; c) recurrence quantification analysis to emphasize the nonlinear dynamics of the exposure levels, respectively the recurring exposure levels dynamics. Similarities and dissimilarities were observed and discussed.

1. Introduction

In the last decades, the development of wireless communication has raised concerns about the potential health effects of electromagnetic field (EMF) on humans [1]. Therefore, the long-term EMF monitoring has become an important scientific research subject, particularly in increased sensitive areas. Valuable information about human exposure levels and their time-variability may be provided by monitoring systems, even certain hidden knowledge from the recorded data series [2]. In diverse areas form many countries were deployed monitoring networks of fixed radio-sensors to measure and assess the EMF levels in the environment, providing foundational database for risk analysis [3].

EMF investigation requires proper measurement and tools, also safety limits and methods for radiofrequency (RF)-EMF exposure assessment, to reduce some possible health risks. In this regard „Guidelines for limiting exposure to EMF (100 kHz to 300 GHz)" – ICNIRP 2020 were republished, by the International Commission for Non-Ionizing Radiation Protection (ICNIRP) [4]. Also, a reference document used in the European Union, is the EU Council Recommendation 1999/519/EC which is based on the reference limits set out by ICNIRP's 1998 guidelines [5].

In June 2021 a preliminary analysis report was released about research activities connected to the health impact of the RF-EMF exposure from mobile networks (mainly base stations). A very high level of uncertainties was reported, in this sense some pro-active initiatives evolved, namely a better-quality of data collection, standard experimentation protocols and cooperation between interdisciplinary scientific research groups [6]. As a solution to this kind of ambiguities and to the increased concern about human exposure to RF-EMFs sources, was the adaptation of a continuous surveillance of electromagnetic areas by the automatic control of the electromagnetic emissions levels. Therefore, a proper method to communicate in real-time the measured levels of EMF was required, in order to provide valuable information on human RF exposure [7].

Up to now, different RF monitoring systems have been deployed in various countries, including Romania, as an important long-term acquisition method of daily exposure levels. Useful guidelines are offered by the ITU-T K.83 recommendation through providing procedures of measurement enabling proper monitoring the EMF on the long term and in different environments [4].

In Romania, the National Authority for Management and Regulation in Communications (ANCOM) is the institution responsible for assessing the human exposure levels to RF-EMF sources. This authority provides an interactive map which can be accessed on the Internet and which presents the EMF levels measured in all cities from Romania compared with the reference levels established by the national regulations [8]. Usually, the fixed sensors are located in the nearest vicinity of congested areas where EMF agglomeration is estimated (i.e. hospitals, educational institutions, public institutions). Using the exposure levels recorded by a continuous monitoring 24/356 (hours/days) recorded at the station of Iasi city for the whole year 2020, the authors of [9] concluded that no exceedance of the safety limits was encountered, if reported against the ones adopted by the national legislation.

The study conducted by Mulugeta et al. [10] presents a comparison between the ambient RF-EMF levels from outdoor areas and public transport in Switzerland in the interval 2014-2021 by using portable equipment carried in a backpack. Using the multilevel regression modelling, they analyzed the distribution and the variability of the ambient RF-EMF levels per micro-area. They concluded that in public places the RF-EMF levels do not present a significant change between 2014 and 2021 despite an 18-fold increase in mobile data transmission and presumed that this behavior is connected with the shifting to the newest mobile communication devices.

In [11] Kurnaz et al. aimed to assess RF-EMF exposure by comparing it with the reference levels set by ICNIRP in one of the largest provinces of Turkey between 2016 and 2018. The findings indicate that short-term RF-EMF levels doubled as a result of the introduction of 4G mobile networks. Also the authors observed that the majority (99.3%) of the measured levels came from the base stations, where devices operating in 3G band contributed the most (31.2%). The research emphasize a different behavior in the phase evaluation, namely an "exponential distribution" of the short-term field levels case and a "Burr distribution" for long-term field strength values.

Djuric et al. [12] presented an approach of the environmental EMF behavior during a five-year period extracted from the data set acquired by a sensor of the Serbian EMF RATEL monitoring network. This study confirmed that the growth of the electric (E)-field strength from year to year is closely related to the installation of the new communication devices in the observed environment.

Chiaraviglio et. al [13] provided a comparative assessment of exposure by provisioning repeated pre- and post-5G services. They observed that the measured traffic became more than double when 5G communications services were activated, reaching levels above 200 Mbps. The research demonstrated the influence of the reconfiguration changes on the environment in order to accommodate 5G panels, also the site conditions and the power received from the main radiation lobes – which were impacted.

Present work aimed to analyze the real-time monitoring data gathered by the national monitoring system of Romania for all months of years 2019 and 2023. One monitoring fixed station was chosen per each of the three cities from Romania considered in the present analysis. Peculiarities of exposure during the examined period of time were emphasized along with a comparison of the measured exposure levels with those established by the international organizations. Based on descriptive statistics and on time-series analyses, some hidden temporal patterns of exposure were highlighted.

2. Materials, measurement system and procedure

Starting from 2015, Romania operates through ANCOM a broadband RF field monitoring network [8]. Currently (beginning of 2024), throughout the country, measurements are conducted exclusively outdoors, using the 150 broadband monitoring stations, measuring in the frequency range of 100 kHz - 7 GHz, located in more than 104 localities across the country. The EMF sensors have been installed for continuous measurement of EMF levels in public areas near concentrations of field sources.

A national map allows visualization of all locations where ANCOM installed EMF sensors together with the measurement results gathered by these sensors along different time intervals. The results are presented to the public in a suggestive graphically manner and the values are compared with reference levels for the general public specified in the national regulations.

The EMF monitoring network comprises of sensors (receiving antennas) placed in fixed locations, specialized software applications, hardware components for storing and processing data transmitted by the sensors, and the necessary communication devices for interconnecting the components, all integrated uniformly within the autonomous EMF measurement system of ANCOM.

(a) (b)

(c)

Figure 1. EMF quad-band monitoring fixed stations used in present study from: (a) Bucharest city, (b) Cluj-Napoca city and (c) Iasi city

The EMF monitoring stations use a triaxial isotropic field probe to measure E-field strength by broadband sensors model NARDA AMB 8057 and NARDA AMB 8059.

Four frequency bands are investigated:

- A wideband covering (100 kHz - 7 GHz) - which includes the bands used in Romania for 5G-FR1 mobile communications);
- Three narrow frequency bands of the mobile telephony downlink spectrum of 2G-4G: (925 – 960) MHz, (1805 – 1880) MHz and (2110 – 2170) MHz.

In this paper, raw data collection and processing originated from only one station placed in each of the following three important cities of Romania: Bucharest (the capital), Cluj-Napoca and Iasi. Complete data records for years 2019 and 2023 constituted the database of our processing. The population of the cities was: a) Bucharest: 1.82 million persons; b) Cluj-Napoca: 0.37 million persons; c) Iasi: 0.32 million persons.

Fig. 1 shows photos of the monitoring stations mounted in each of the investigated locations from the three cities.

Three types of computations were made, using the database of E-field strength values for the three locations and two different years of analysis: a) descriptive statistics revealing the boxplot distributions of the average E-field strengths; b) Poincare plots of the time series of data, with the aim of extracting the trends of the short- and long-term variabilities of the exposure levels [14]; c) recurrence quantification analysis (RQA) which allows nonlinear data analysis for the investigation of dynamical systems; in the present case, RQA may reveal determinism of the trends.

3. Results and discussion

The mean, median and interquartile ranges (IQR) of the E-field strengths measured during 12 months corresponding to 2019 and comparatively to 2023 by the monitoring stations are shown in the boxplots of Fig. 2. Four different frequency bands were analyzed: a) the whole bandwidth (100 kHz – 7 GHz); b) the mobile downlink band (920-965) MHz; c) the mobile downlink band (1805-1880) MHz; d) the mobile downlink band (2110-2170) MHz. The measured data were exported over course of the 2019 year and 2023 year. None of the recorded values exceeded the ICNIRP reference levels for the general public safety. Interquartile ranges represent a valuable tool to express the median variability and give us information about the range of the middle half of the measured data.

The largest IQR in which E-field strengths spread was observed for Bucharest in 2019 year. Also for Cluj-Napoca city in both years' cases, there are large IQR intervals. As we can observe in Figure 2, there are some specific cases where measured data are marked by outliers (rhombus), indicating that E-field are being too far from the central values to be reasonable; also this emphasized the fact that here the variability of the field is the highest. In terms of exposure, we noticed that in 2019 year, higher exposure was reported in Bucharest than in 2023, in the chosen location. Otherwise in 2023 in Cluj-Napoca city, the exposure level exceeded the one measured in 2019. This type of behavior may derive from the number of RF emitters in use in the respective years. Also, a particular case is represented by Iasi city where we observed that for the three frequency sub-bands, the highest E-field strengths were recorded in 2023, along with a lower variability of the measured values.

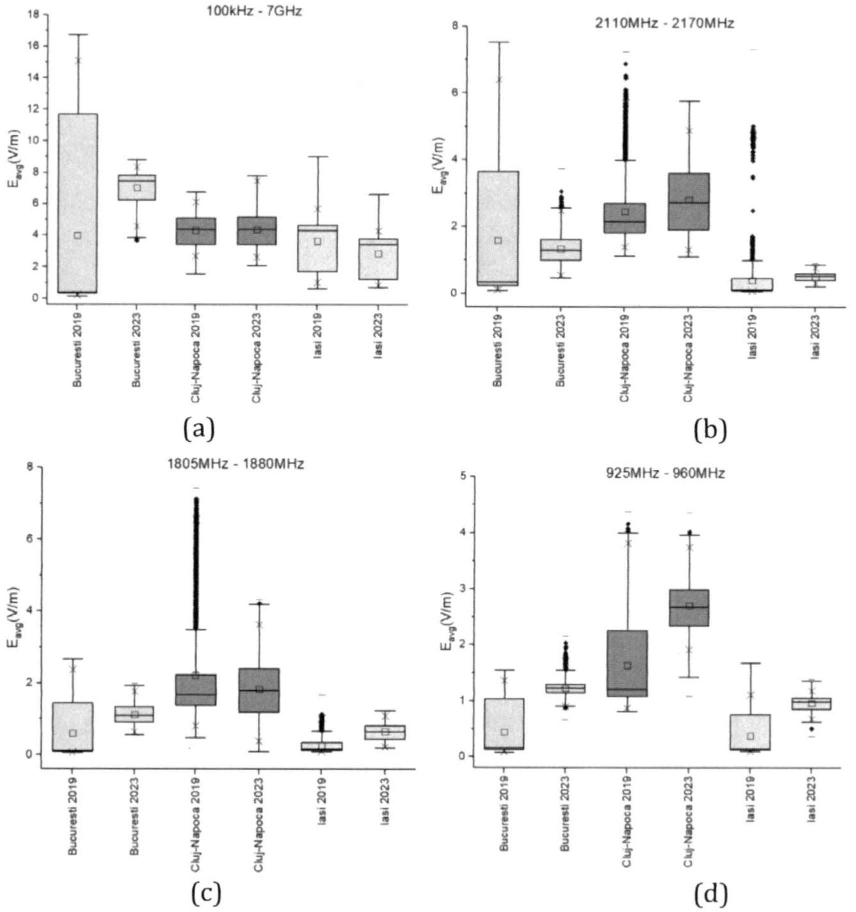

Figure 2. Boxplots of the distributions of the average field levels (E_{avg}) with outliers, in the three cities/locations analyzed in 2019 and 2023: (a) wide bandwidth; (b) – (d) narrow bandwidths corresponding to mobile downlink emissions (base stations)

Fig. 3 exemplifies the characteristic Poincare plots of the time-distribution of E_{avg}-field strengths measured in the 100 kHz – 7 GHz bandwidth corresponding to the three cities in 2019 and 2023. We noticed that for all the cities it resulted the same regularity: a smaller short-term variability of exposure was present than the long-term variability. This may be due to the continuous change of the electromagnetic environment, by introducing new RF sources (including 5G). For Iasi city we observed a similar field variability and a higher entropy of the time series of data also proved by the RQA variables Also a great multitude of data points were distributed in the same pattern along SD2 axis (long-term variability indicator). The higher variability of the signals is also proved by the majority of data points being spread outside of the entropy. In Cluj-Napoca in 2019 a higher variability can be observed compared with the one in 2023. For

Bucharest city, in 2023 there are various points which are spreaded outside the line of the ellipse, fact correlated with the presence of the multiple outliers from the boxplot chart.

Fig. 4 shows the Poincare plots representations corresponding to the three cities, of the time-series of E_{avg}-field strength measured in the (925 – 960) MHz bandwidth for both years. In case of Bucharest city for 2019 the time-series data varies in a large interval, aspect also met in the boxplot chart. In 2023 year the recurrent data points are mostly distributed outside the edges of the Poincare ellipse, these data set are being represented by the outliers from the boxplot graph. Here, for all cities, in 2023 year we noticed a higher entropy which reflected a slightly increased complexity of the deterministic structure in the time-series of data.

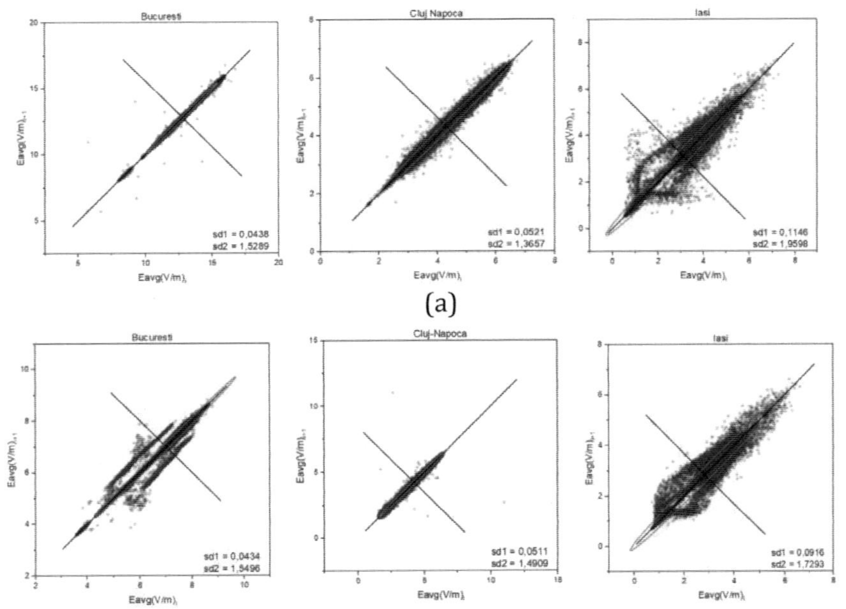

(b)

Figure 3. Poincare maps of E_{avg}-field strengths for the three cities in the wideband of (100 kHz – 7 GHz) for: (a) 2019 year and (b) 2023 year

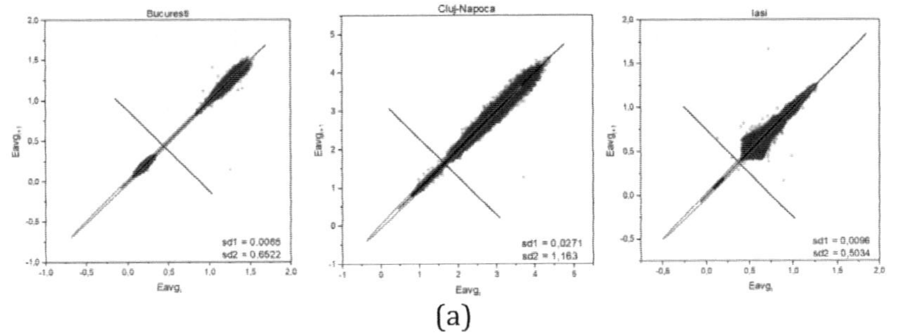

(a)

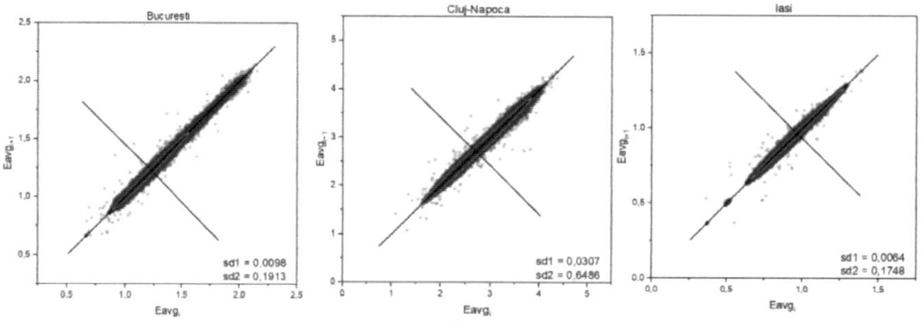

(b)

Figure 4. Poincare maps of E_{avg}-field strengths for the three cities in the (925 – 960 MHz) bandwidth for: (a) 2019 year and (b) 2023 year

Fig. 5 show the Poincare graphs for all cities in the (1805 – 1880 MHz) bandwidth, corresponding to 2019 and 2023 years. We noticed a significant behavior in the recurrence Poincare plots in 2023, more exactly the signals tend to have a certain scattering pattern for all 3 frequency sub-bands and all cities, compared to those in 2019 year. Therefore the time-variabilities are dependent on the frequency band and the level of usage of mobile communication in some certain periods of time. Here, for Cluj-Napoca city in 2019 we observed a higher variability of the signal, proved by the great number of data points which are outside of the Poincare ellipse and the outliers from the boxplot chart.

Also, in Fig. 6 are illustrated the Poincare plots for (2110 – 2170) MHz bandwidth, 2019/2023 years and all the three cities. In this case we observed a different variation of the field strength, the majority of recurrence points being outside the ellipse and at a greater distance from it, aspect strengthened by the RQA parameters and the presence of the outliers. In addition, according to the usage of a high-frequency bands (upper-bands) of mobile communication devices, we noticed signals with a higher variability and larger ratio of SD1/SD2 axes (very different variabilities on the short- versus long-term).

Based on the short-term variability (SD1) and long-term variability (SD2) axes of the Poincare graphs, we calculated the areas of the ellipses (Fig. 7), as they illustrate the total variability of the time-series. Here we observed that for Bucharest and Iasi cities, the total variabilities of the emissions were higher in 2019 for all frequency bands than in 2023. Interesting to note is that in the (925-960) MHz and (1805-1880) MHz bands, reduced values of the total variability were present, indicating that the base stations emitting in these bands in the investigated locations didn't suffer major changes. In the bandwidth (2110-2170) MHz belonging to 4G mobile communications, the most significant changes appeared in Bucharest, where the variability increased almost 10 times from 2019 to 2023, while in Iasi it decreased 3.2 times in the same interval.

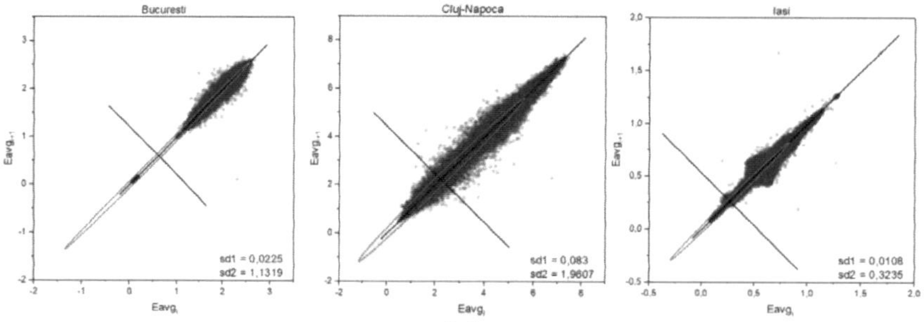

(a)

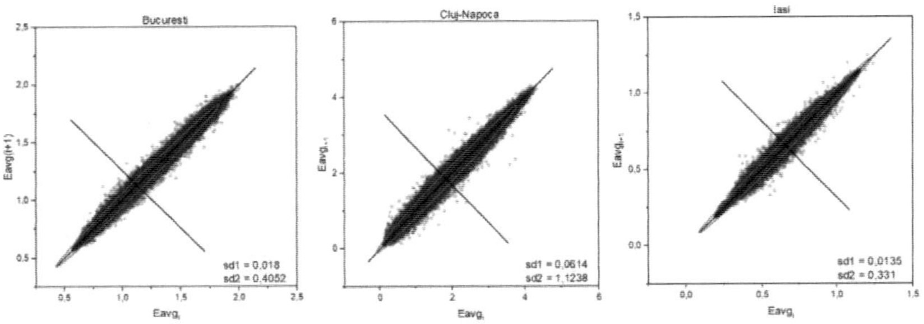

(b)

Figure 5. Poincare maps of E_{avg}-field strengths for the three cities in the (1805 - 1880 MHz) bandwidth for: (a) 2019 year and (b) 2023 year

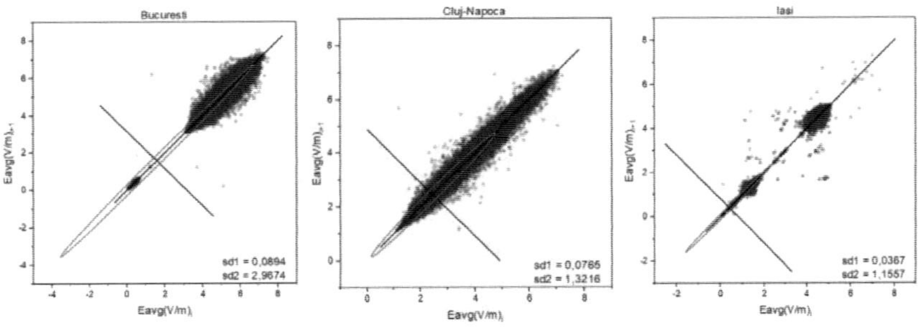

(a)

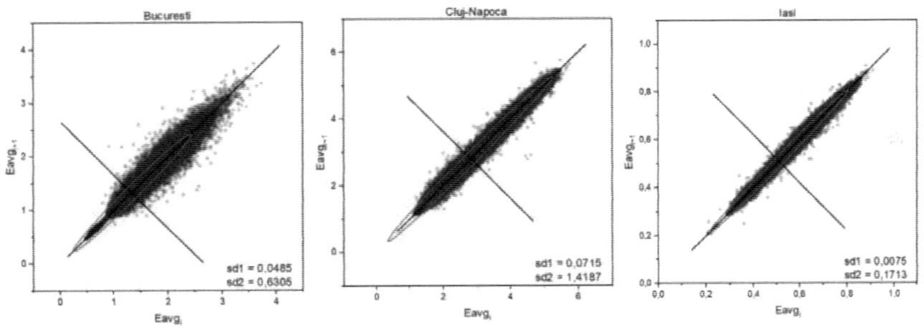

(b)

Figure 6. Poincare maps of E_{avg}-field strengths for the three cities in the (2110 - 2170 MHz) bandwidth for: (a) 2019 year and (b) 2023 year

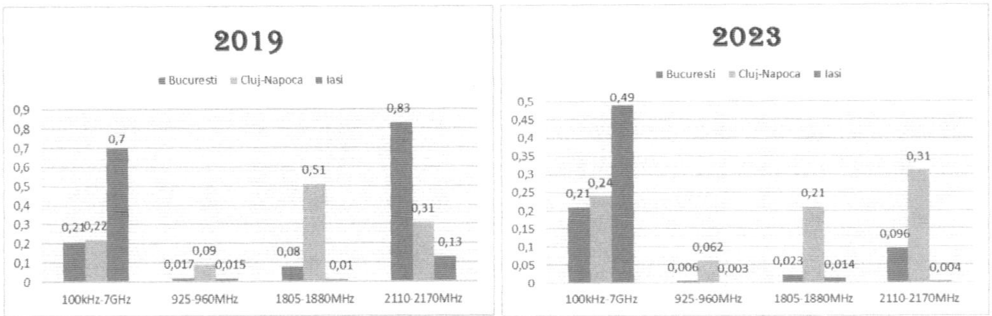

(a) (b)

Figure 7. Total variability of exposure expressed by the areas of Poincare maps of E_{avg}-field strengths for all cities in case of whole spectrum and restricted bandwidths: (a) 2019 year and (b) 2023 year

IOP Conf. Series: Materials Science and Engineering **1320** (2024) 012009 doi:10.1088/1757-899X/1320/1/012009

Table 1 shows the RQA parameter values for the wide-band and for the 3 frequency sub-bands in case of Bucharest (B.), Cluj-Napoca (CJ.) and Iasi (IS.) cities for 2019 year and Table 2 shows the RQA parameters for the same locations but for 2023 year. These parameters provide more information about the quantification of the number and duration of recurrences specific to a dynamic system.

First of all, we observe that in Iasi city in both years, the determinism parameter (DET) indicate the existence of a deterministic system, i.e. sequences of the same exposure levels repeating at different time intervals. Also, higher DET values are present in 2019 year case, which indicates a "higher" determinism than in 2023 year case. All the other cities present reduced DET parameter values which implies a low degree of determinism of the exposure level.

A greater degree of determinism is strictly correlated with a higher value of recurrence rate (RR), as we can see in case of Iasi city, in this sense a high density of recurrence points is indicated. For Bucharest and Cluj-Napoca cities the RR parameter has values which not exceeded 1, thus this aspect strengthens and validates the analyzed behavior expressed by the DET parameter.

The Laminarity (LAM), trapping time (TT) and longest diagonal line (L_{max}) parameters of the annual time series quantify the time interval during which the exposure does not modify appreciably. When the values of LAM parameter decreases, more singular recurrence points are present compared to the vertical lines from graphical representations. The negative values of the Trend parameter indicate the exposure level departing from the initial state with a continuous changing. Here there are some particularities, as we can observe from Table 1 and Table 2, namely the 0 values of Trend parameter which stands for a linear trend where the exposure level does not follow a similar rule of variation.

Table 1. The RQA parameters of the time series of exposimetric data, in case of whole spectrum and restricted bandwidths for 2019 year

RQA variables	100 kHz-7 GHz			925 – 960 MHz			1805 - 1880 MHz			2110 – 2170 MHz		
	B.	CJ.	IS.	B.	CJ.	IS.	B.	CJ.	IS.	B.	CJ.	IS.
Determinism (DET)	0	0	75	-1	0	10	-1	0	15	-1	-1	9
Reccurence rate (RR)	0.06	0.06	7.4	0	0.02	1.6	0	0.02	3	0	0	0.3
Laminarity (LAM)	-1	0	59	-1	0	7	-1	0	5	-1	-1	2
Trapping time (TT)	-1	-1	15	-1	-1	-1	-1	-1	-1	-1	-1	-1
Entropy (ENTR)	-1	-1	3.4	-1	-1	-1	-1	-1	-1	-1	-1	-1
Trend	-0.6	-2.2	-250	0	-0.7	-47	0	-0.68	-35	0	0	-11
Longest diagonal line (L_{max})	-1	-1	32	-1	-1	-1	0	-1	-1	-1	-1	-1

Table 2. The RQA parameters of the time series of exposimetric data, in case of whole spectrum and restricted bandwidths for 2023 year

RQA variables	100 kHz-7 GHz			925 – 960 MHz			1805 - 1880 MHz			2110 – 2170 MHz		
	B.	CJ.	IS.	B.	CJ.	IS.	B.	CJ.	IS.	B.	CJ.	IS.
Determinism (DET)	0	-1	19	0	0	4	0	0	10	0	0	3
Reccurence rate (RR)	0.5	0	7.9	0.9	0.02	3.57	0.18	0.02	1	0.3	0.02	5.9
Laminarity (LAM)	0	-1	39	0	0	5	0	-1	3	0	0	1
Trapping time (TT)	-1	-1	14	-1	-1	-1	-1	-1	-1	-1	-1	11.5
Entropy (ENTR)	-1	-1	2	-1	-1	-1	-1	-1	-1	-1	-1	1
Trend	-3.7	-1.4	-11	-61.8	-0.6	-37.4	-1.18	-0.7	-1.4	-11	-0.7	-174
Longest diagonal line (L_{max})	-1	-1	30	-1	-1	-1	-1.2	-1	-1	-1	-1	11

4. Conclusions

The ever-growing number of RF emitting sources in the environment, including the cellular communications technologies, are continuously increasing the EMF level in all environments. From this perspective, the EMF monitoring has become a paramount activity in order to manage the possible health risks of the population. Present paper focuses on demonstrating the usefulness of analyzing the raw data provided by the fixed monitoring stations of ANCOM in Romania, which entered into function in 2015. In the current approach we analyzed the evolution of the exposure levels and its variability during two complete-data years separated by a 4-years interval, namely 2019 and 2023, in one location chosen in each of three important cities in Romania.

All the recorded levels were situated below the reference levels established in the ICNIRP guidelines for the safety of the population.

By illustrating in the boxplot charts the average E-field strengths distributions along 12 months, different features could be established : a) the largest inter-quartile values distribution of emitted E-field strengths were observed for Bucharest city in 2019 year; b) the presence of the outliers values indicate a significant deviation of the E-field from the central values and the highest variability of the field (maybe due to artefacts); c) the exposure levels depend on the traffic recorded during the measurement time; d) in Cluj Napoca and Iasi, a growth trend of the E-field level was demonstrated from 2019 to 2023, probably a consequence of new communication sources installed in the area of the observed environment.

A clear view of the time-variability of exposure is given by the Poincare plots which are based on the time-series analysis, so that different peculiarities could be emphasized: a) a smaller short-term variability of exposure than long-term variability was present in all situations; b) higher variability of the exposure was proved by the majority of points spread outside the entropy, behavior correlated with the outliers presence revealed by the boxplot charts; c) for all analyzed bandwidths in Bucharest city, the time-series data varied in the largest interval; d) higher entropy of the exposure level evolved, reflecting a slightly increased complexity of the deterministic structure inside the time-series; e) similar scattering pattern of RF levels resulted for all 3 frequency sub-bands and for all cities in 2023; f) usage of mobile communication devices that operate in the high-frequency bands (including 5G) imply higher variability of the exposure levels and larger short- and long-term variabilities, as indicated by the SD1 and SD2 axes of the Poincare ellipses. The computed Poincare plot areas conducted to the conclusion that the highest variabilities of the field levels were encountered in 2019 in Bucharest and Iasi cities.

Overall, the large database collected by entering in use of all the fixed RF monitoring stations of the ANCOM network offer excellent starting points for specific and more complex analyses of the dynamics and trends of the human exposure levels in various locations and times.

References

[1] European Commission. Joint Research Centre. Electromagnetic Emissions from Mobile and Potential Effect on Health: Preliminary Study.; Publications Office: LU, 2021.

[2] Bhatt, C.R., Henderson, S., Brzozek, C., Benke, G., 2022. Instruments to measure environmental and personal radiofrequency-electromagnetic field exposures: an update. Phys Eng Sci Med 45, 687–704.

[3] Jalilian, H., Eeftens, M., Ziaei, M., Röösli, M., 2019. Public exposure to radiofrequency electromagnetic fields in everyday microenvironments: An updated systematic review for Europe. Environmental Research 176, 108517.

[4] ITU-T, Monitoring of electromagnetic fields, Series K, Supplement 83, 2022. Available online: https://www.itu.int/rec/T-REC-K.83-202201-I/en.

[5] 1999/519/EC: Council Recommendation of 12 July 1999 on the limitation of exposure of the general public to electromagnetic fields (0 Hz to 300 GHz).

[6] European Commission. Joint Research Centre., 2021. Electromagnetic emissions from mobile networks and potential effect on health: preliminary study. Publications Office, LU.

[7] Apostolidis, C., Manassas, A., Iakovidis, S., Samaras, T., 2022. Electromagnetic Fields in the Environment: An Update of Monitoring Networks in Europe, 1-3.

[8] ANCOM, Monitor EMF. Available online: https://monitor-emf.ro/map.

[9] Ursachianu, M., Bejenaru, O., Lazarescu, C., Salceanu, A., Păuleţ, M., 2021. The Assessment of Human Exposure in Iasi-City using Data Provided by The National Autonomous RF-EMF Monitoring System throughout 2020, 225-230.

[10] Loizeau, N., Zahner, M., Schindler, J., Stephan, C., Fröhlich, J., Gugler, M., Ziegler, T., Röösli, M., 2023. Comparison of ambient radiofrequency electromagnetic field (RF-EMF) levels in outdoor areas and public transport in Switzerland in 2014 and 2021. Environmental Research 237, 116921.

[11] Kurnaz, C., Mutlu, M., 2020. Comprehensive radiofrequency electromagnetic field measurements and assessments: a city center example. Environmental Monitoring and Assessment 192.

[12] N. Djuric, T. Skoric, D. Kljajic, V. Otasevic, S. Djuric, 2023. A Hidden Knowledge in Long-Term EMF Monitoring of EMF RATEL Monitoring Network, in: 2023 Photonics & Electromagnetics Research Symposium (PIERS). 2023 Photonics & Electromagnetics Research Symposium (PIERS), 1856–1861.

[13] Chiaraviglio, L., Bartoletti, S., Blefari-Melazzi, N., Lodovisi, C., Moretti, A., Zampognaro, F., Alouini, M.-S., 2023. Measuring EMF and Throughput Before and After 5G Service Activation in a Residential Area. IEEE Open Journal of the Communications Society PP 4, 1179–1195.
[14] Miclaus, S., Sarbu, A., Bechet P., 2021. Using Poincare Plots for Feature Extraction of the Dynamics of Electromagnetic Field Exposures when Using Different Protocols of Wi-Fi Communications. Proceedings of the 2021 8th International Conference on Wireless Communication and Sensor Networks (icWCSN '21). Association for Computing Machinery, New York, NY, USA, 32–38.

IOP Conf. Series: Materials Science and Engineering **1320** (2024) 012010 doi:10.1088/1757-899X/1320/1/012010

An innovative mixed reality approach for Robotics Surgery

Gabriela Rus [1], Nadim Al Hajjar [2], Ionut Zima[1], Calin Vaida[1], Corina Radu[3], Damien Chablat [1,4], Andra Ciocan[2] and Doina Pîslă[1,5*]

[1] CESTER, Research Center for Industrial Robots Simulation and Testing, Technical University of Cluj-Napoca, 400641 Cluj-Napoca, Romania
[2] Department of Surgery, "Iuliu Hatieganu" University of Medicine and Pharmacy, 400347 Cluj-Napoca, Romania
[3] Department of Internal Medicine, "Iuliu Hatieganu" University of Medicine and Pharmacy, 400347 Cluj-Napoca, Romania
[4] Nantes Université, École Centrale Nantes, CNRS, LS2N, UMR 6004, F-44000 Nantes, France
[5] Technical Sciences Academy of Romania, 26 Dacia Blvd, 030167 Bucharest, Romania

*E-mail: doina.pisla@mep.utcluj.ro

Abstract. Robotic-assisted procedures offer numerous advantages over traditional approaches, including improved dexterity, reduced fatigue, minimized trauma, and superior outcomes. However, the main challenge of these systems remains the poor visualization and perception of the surgical field. The goal of this paper is to provide an innovative approach concerning an application able to improve the surgical procedures offering assistance in both preplanning and intraoperative steps of the surgery. The system has been designed to offer a better understanding of the patient through techniques that provide medical images visualization, 3D anatomical structures perception and robotic planning. The application was designed to be intuitive and user friendly, providing an augmented reality experience through the Hololens 2 device. It was tested in laboratory conditions, yielding positive results.

1. Introduction

Procedures involving critical organs like the pancreas and esophagus have advanced significantly in the field of medical robots. Robotic technologies have transformed surgical procedures on these essential organs, giving surgeons greater control and precision as they navigate complex anatomical structures. Robotic platforms provide greater dexterity for surgeries involving the pancreas, which are sometimes complex due to the organ's location and function. Similarly, the accuracy and minimally invasiveness of robotic-assisted techniques have revolutionized surgeries involving the esophagus, with its intricate network of tissues and close proximity to important structures. Consequently, patients undergoing esophageal and pancreatic procedures had better postoperative results, shorter recovery periods, and less stress, highlighting the critical role that medical robotics plays in these types of procedures. [1-2].

The favourable results provided by assisted robotic surgery have led to widespread acceptance of robots in these procedures, with recent decades witnessing accelerated development of many structures designed for this purpose [3-5]. These structures have proven particularly suitable for procedures targeting organs located in hard-to-access areas such as the pancreas or oesophagus. However, despite the undeniable advantages, robots used for these types of procedures bring with them some limitations, such as appropriate visualization of the surgical field. In order to compensate for this limitation, devices have begun to be used to provide insights about the patient during the procedure, augmenting the surgeon's environment.

Within the domain of medical robotics Augmented Reality (AR) techniques find applications in rehabilitation, surgery, and medical assistance. Within surgical contexts, AR is integrated across various stages, from preoperative planning [6] to intraoperative guidance [7] and telesurgery [8].

In [9] and [10], the authors investigated the application of an AR interface to optimize trocar placement, reducing the risk of robot arm collisions in laparoscopic procedures and facilitating robot configuration initialization in teleoperated robot-assisted surgeries with the da Vinci system.

Fu et al. [11] developed an AR-assisted robotic learning framework tailored for MIS tasks, utilizing an AR interface through the HoloLens 2 OST-HMD to validate the configuration of a serial redundant robot. To establish a reliable trajectory for replication by the robotic system in minimally invasive surgery scenarios, the Gaussian mixture model (GMM) and Gaussian mixture regression (GMR) were employed to encode multiple trajectories demonstrated by humans in this study.

Despite advancements in AR applications for surgery, there's a noticeable gap in comprehensive solutions covering both preplanning and intraoperative phases. Furthermore, just few applications address simulations and the integration of robotic systems. Bridging this gap requires collaborative efforts to design AR solutions benefiting surgical teams throughout the entire procedure.

The aim of the paper is to present an application designed to improve surgical procedures by offering insights into the patient during both preplanning and intraoperative phases. The proposed system is designed to: (i) provide an adequate overview of robotic systems integration in the operating room, assisting both surgeons and engineers in finding suitable solutions for Robotic Assisted Surgery (RAS), (ii) facilitate access to medical images and anatomical details of the patient during surgery to improve real-time guidance, and (iii) offer a superior method for pre-planning through features such as 3D anatomical structure visualization, the ability to highlight abnormalities, and visualization of volume rendering to anticipate potential challenges. Following the introduction in section 1, section 2 provides the methodology for developing the application, including a description of the structural architecture. Section 3 presents the results, covering both the design and performance of the application. Finally, section 4 encapsulates the conclusions drawn from this study.

2. Methodology

2.1 Description of the architectural structure

The application has been developed in Unity, using the MRTK (Mixed Reality Toolkit) packages which provides a set of tools and components used for developing mixed reality applications across a variety of platforms, primarily focusing on Microsoft HoloLens and Windows Mixed Reality devices.

The inputs for the application, including CT scans, 3D organ reconstructions, and the CAD models of the robotic structures, have been provided by other software tools such as Slicer 3D. Volume data has been exported in Nearly Raw Raster Data (NRRD) format, while 3D reconstructions are in OBJ format. Additionally, Siemens NX was used to export the robotic structures in STEP format, as illustrated in Figure 1.

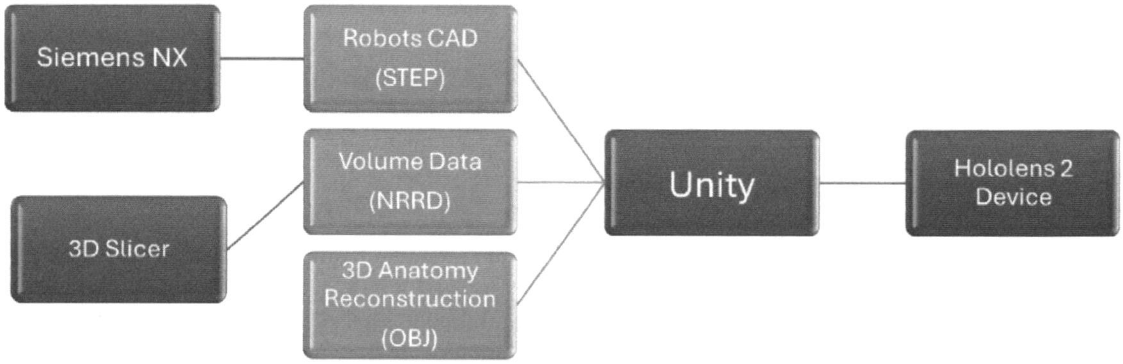

Figure 1. Architectural structure of the proposed application

2.2 Description of the targeted procedures

For a better understanding of the importance of this application in surgery, an overview of the complex task imposed by pancreatic and oesophagus surgery is required.

A) Minimally invasive techniques for esophagectomy

Esophagectomy, essential for the treatment of malignant and benign diseases, relies on surgical team's expertise and tumor location. Minimally invasive techniques, including robotics, offer advantages but pose challenges such as high risk of anastomotic leaks. The ongoing Robotic-Assisted Minimally Invasive Esophagectomy (RAMIE) trial evaluates robotic-assisted procedures' efficacy, showing promising early results. These procedures can be approached through three methods.

1. Transhiatal esophagectomy

This procedure is specific for tumors in the upper or middle oesophagus.

The procedure is performed as follows: the patient is positioned supine with a steep reverse Trendelenburg tilt, and trocars are placed for robotic assistance and liver retraction. The robot is positioned for optimal tumor visualization. Dissection begins along the greater curvature of the stomach, followed by division of the pars flaccida and creation of a window in the lesser curvature. The left gastric artery is isolated and divided, and the oesophagus is mobilized transhiatally all the way to the cervical region. A cervicectomy is performed. After examining perfusion with ICG, the gastric conduit is anastomosed to the specimen, and the robot is undocked.

2. Robotic Ivor Lewis esophagectomy

The patient is positioned supine with a reverse Trendelenburg tilt. Trocars are placed for robotic assistance, assistant support, and liver retraction. Ligaments are divided, arteries and veins are prepared, and an omental flap is created. Mediastinal and nodal dissections is performed. The gastric conduit is created, mobilized in the thorax, where the anastomosis will be done. A jejunostomy tube is inserted at the end and incisions are closed.

3. Robotic McKeown esophagectomy

The patient is positioned in a left lateral decubitus, semi-prone, with a 45° anterior tilt. Trocars are placed for robotic assistance and the assistant. The right lung is deflated and retracted after dividing the inferior pulmonary ligament, and the oesophagus is mobilized up after the azygos cross is stapled. Periesophageal and mediastinal lymphadenectomy is performed. The oesophagus is dissected and resected, and the specimen is retrieved. The abdominal time implies greater and lesser curvature of the stomach dissection. With further mobilization of the gastric conduct in the cervical region, where the anastomosis will be done. After closure of incisions, the robot is undocked [12].

B) Minimally invasive techniques for pancreatectomy

Robotic surgery is increasingly utilized for pancreatic lesions and malignancies providing advantages such reduced post-operative pain, precise manoeuvrability (allows for cutting and suturing at angles not possible for humans' hands) safety and feasibility [13]. The most standard procedures are:

1. **Cephalic pancreatoduodenectomy (Whipple procedure)**

The procedure, performed with robotic assistance involves positioning the patient supine and inserting trocars for camera and robotic arms. The surgeon mobilizes and transects the head of the pancreas, entire duodenum, gallbladder with the common bile duct, with or without pylorus preservation, meticulously preserving vital surrounding structures such as the portal and superior mesenteric vein and artery. Lymphadenectomy is mandatory around the liver hilum and celiac trunk. Then, reconstruction time involves anastomoses to restore gastrointestinal continuity and pancreatic function: Wirsungo-jejunostomy or pancreato-gastrostomy, hepatico-jejunostomy and gastro-jejunostomy [14].

2. **Distal pancreatectomy with or without spleen preservation**

The patient is positioned supine, and trocars are inserted for camera and robotic arms. The surgeon dissects and resects the distal pancreas, with peripancreatic lymphadenectomy with spleen preservation or en-bloc with splenectomy ensuring no iatrogenic trauma of adjacent blood vessels and surrounding organs. Proximal pancreatic stump is carefully stapled to avoid pancreatic leakage and fistula.

2.3 Robotics systems for minimally invasive surgery

For the simulator module of the current application, the UR collaborative robot and a new spherical robot-developed for MIS procedures were used.

The UR5 has 6 degrees-of freedom (DOF), providing a high degree of flexibility and manoeuvrability [15]. The robot's collaborative features enable it to work alongside surgeons, providing steady support and executing predefined trajectories with high repeatability. This collaboration enhances surgical accuracy while minimizing the risk of tissue damage [16].

The spherical robot developed for MIS is presented in Figure 3. The system provides 3 DOF allowing for one translational and one rotational movement used to achieve the desired position under the insertion point, and another translational movement for insertion/extraction of the instrument from the patient's body. Additionally, a counterweight system dynamically counterbalances the instrument load based on gravity compensation, providing structural stiffness. This structure facilitates seamless interchangeability between instruments and the endoscopic camera, alongside a small footprint that allows for easy integration in the operating room [17].

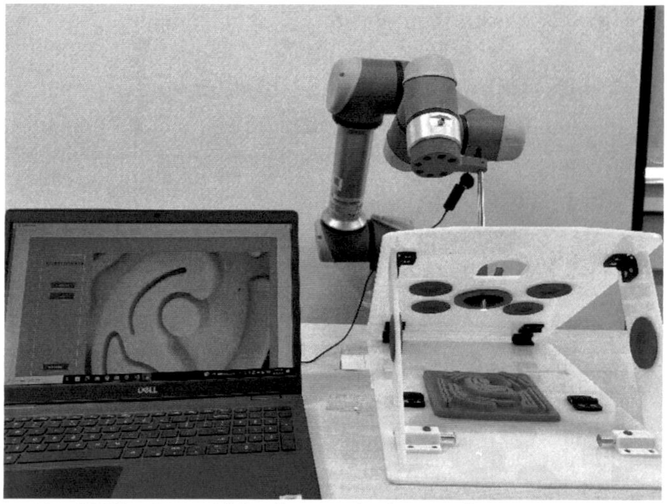

Figure 2. The collaborative UR5 robot within the surgical procedure

Figure 3. Spherical robot within the surgical procedure

2.4 Volume Data

CT data, being in Digital Imaging and Communications in Medicine (DICOM) format, has been imported to Slicer 3D program in order to convert the data and to export it in NRRD format, the NRRD offering many advantages such as **simplicity** - NRRD files are typically simpler and more lightweight compared to DICOM files, **ease of Use** - NRRD files are easier to read and manipulate programmatically compared to DICOM files and **reduced overhead** - NRRD files typically have less overhead compared to DICOM files, which can be advantageous in scenarios where storage space or network bandwidth is limited. Considering that Hololens has limitation related to processing and memory, the NRRD format was considered the most suitable. The NRRD file exported was used for both volume rendering and slicers visualizing.

When the NRRD file has been loaded and the raw data in it converted, the next step was to apply the maximum intensity projection (MIP) algorithm. This method is used to calculate the highest intensity value for each viewing ray through a volume, which is then projected onto a 2D display. Consequently, bones or contrast-enhanced regions become more clearly visible while other parts of the image remain unaltered, and this leads to a better final image. After using MIP algorithms on an image volume, voxel values are transformed into colors and opacities through transfer functions that define shades and opacity of materials thus providing distinction between different tissues and improving visual contrast as well. At this point the rendered volume could be viewed as comprehensive 3D representation of all data stored in NRRD file with emphasis on high density structures since these have been facilitated by MIP technique and color-coded basing on transfer function settings.

2.5 3D Anatomy Reconstruction
3D Slicer was also used in the 3D reconstruction of anatomical structures. DICOM files were utilized to perform semantic segmentation of the anatomical regions of interest. The segmentation process was automated using the Total Segmentator module provided by the software [18]. Subsequently, the segmented organs were imported into Unity in OBJ format.

2.6 Hololens 2 device
HoloLens 2 is designed to merge a virtual world (holograms) with the real environment of the user, creating a mixed reality (MR) experience. This device equipped with a high-definition wave guide display system, coupled with an in-house built holographic processing unit (HPU), for full immersion into visuals and rendering at the speed of thought. The smart glasses are designed to facilitate natural user interaction through improved hand and eye tracking; making it excellent for mapping out 3D environments on which any virtual object could be placed accurately [19].

3. Results

3.1 Functionalities
The application has developed to have a user-friendly design and intuitive functionality architecture. Thus, the design for this application has been thought to be intuitive (as can be seen in Figure 4), using the MRTK packages which contribute with sliders, buttons, text boxes, etc.

Considering the complexity of the robotic assisted surgery, the application has been developed to offer help both in preplanning and intraoperative phase, providing useful insights about the patient. The functionalities include:

Slicer visualization - This functionality allows surgeons to visualize slicers of the CT scans in Hololens 2 device, enhancing spatial understanding and aiding in decision-making. Real-time access to CT scans during surgery enables precise navigation. Furthermore, holographic overlays align virtual structures with patient anatomy, reducing errors. As can be seen in Figure 5 the slicers can be manipulated offering a smooth translation between them.

Anatomical structure visualization – functionality enables the medical staff to plan surgical approaches, anticipates potential challenges, and optimizes procedures for better outcomes in preplanning phase and enhance spatial awareness, facilitating precise surgical maneuvers during surgery. As can be seen in Figure 6 this functionality allows to manipulate the anatomical structure of the body separately.

Volume rendering – offers comprehensive, contrast-enhanced, and easily interpretable visualizations of anatomical structures, the surgeon being able to identify critical landmarks and navigate complex surgical anatomy. In Figure 7 it can be seen how this is particularly helpful for complex procedures, such as esophagectomy, in which the surgeon needs to remove the tumor using an intercostal approach [20].

Underlying abnormalities – can be used for detecting and assessing abnormalities for appropriate diagnosis and treatment planning, as can be seen in Figure 7 a). The application recognizing a gest (tripod grip) and starts draw based on the position of the hand.

Simulation – this functionality provides an adequate environment to plan and observe the entire surgical procedure. This component of the application can be very useful especially in robotic assisted surgery, considering the complexity of procedures and the idea that the robots should be integrated in a specific operating room. As demonstrated in Figure 8, the simulation functionality not only offers an environment for planning and observing surgical procedures but also facilitates the visualization of the integration of multiple robots for collaborative medical applications. This feature is particularly valuable in the context of robotic-assisted surgery, where intricate procedures necessitate careful planning and coordination within specialized operating rooms.

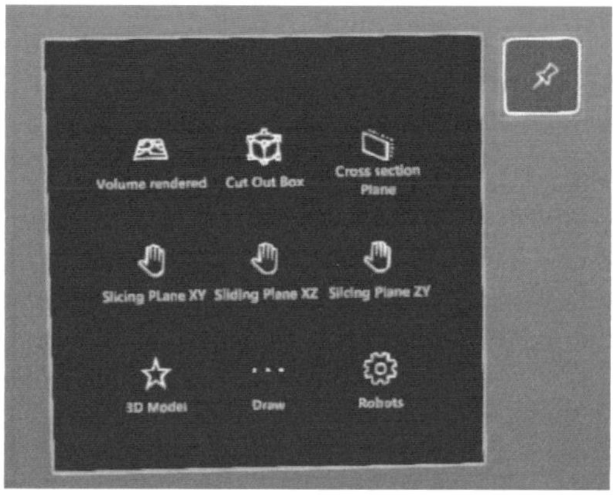

Figure 4. User Interface visualized from Hololens. The Interface includes buttons for volume rendering, Cut Out Box- direct cut in volume rendering, cross section plane – create a section in the plane, slicers visualization in all 3 planes, visualization of 3D anatomical structures, drawing module, simulations and visualization of surgical robots.

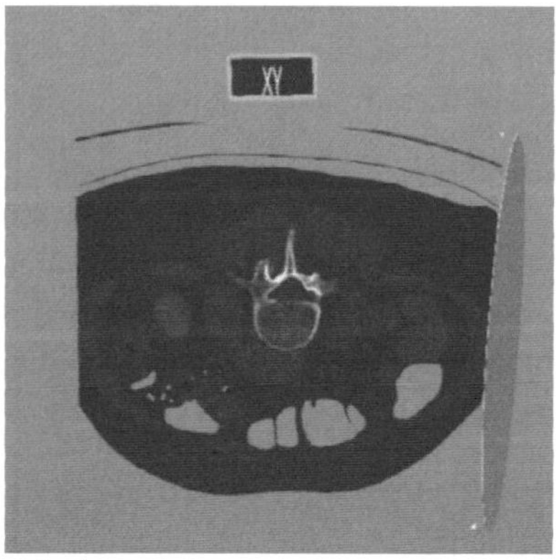

a) Coronal plane

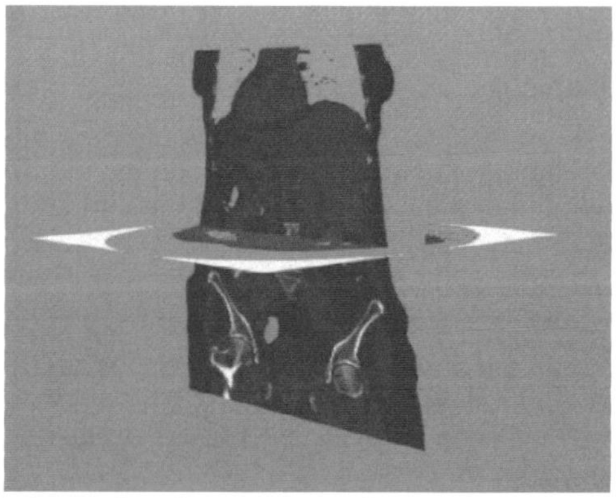

b) Sagittal and Transverse plane

Figure 5. Slicer visualization

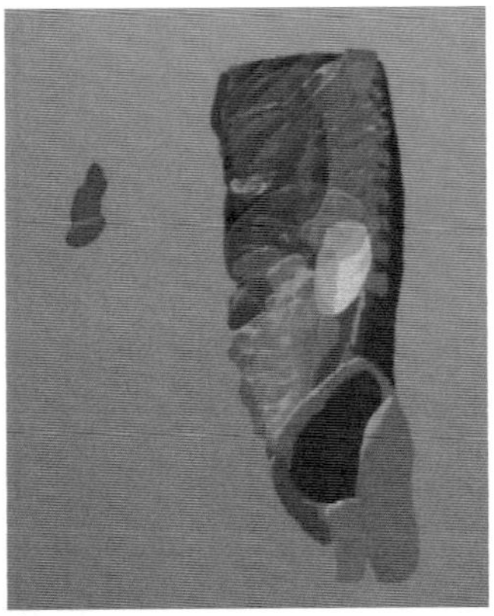

a) Lateral view

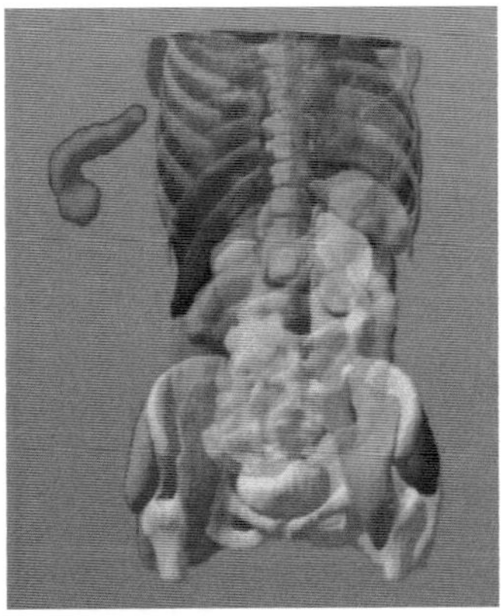

b) Frontal view

Figure 6. Anatomical structure visualization

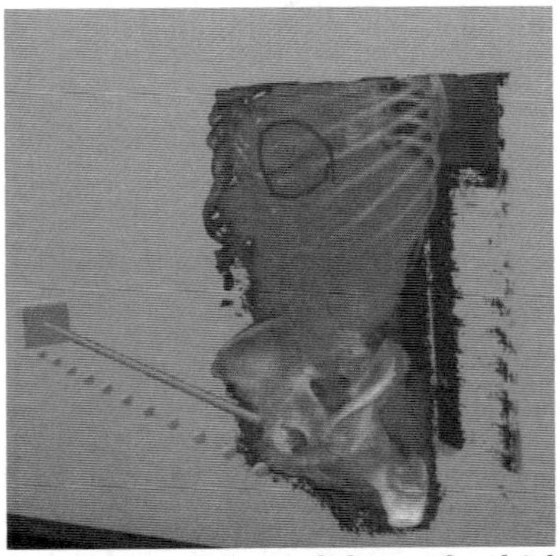

a) Functionality provides a slider with which the visible value range can be changed. Within this functionality the surgeon is able to see the desired range

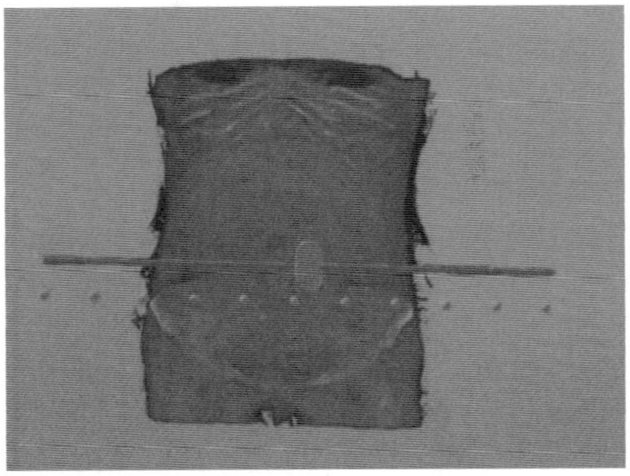

b) The ranges of values for the rendering changed

Figure 7. Visualization of Volume rendering

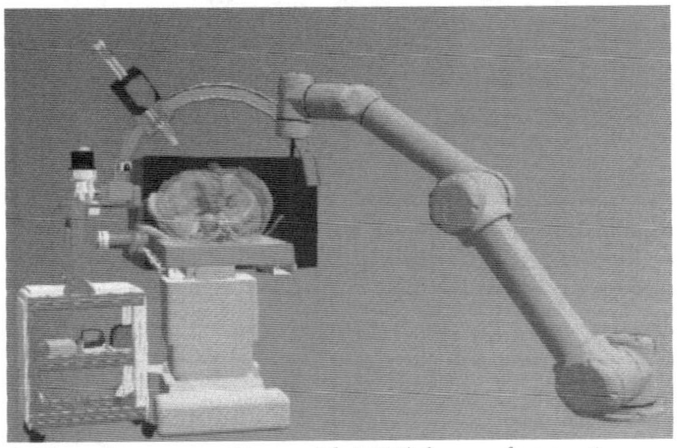

a) Planning operation with UR5 (as endoscopic holder) and the spherical arm (with the surgical instrument), targeting the pancreas.

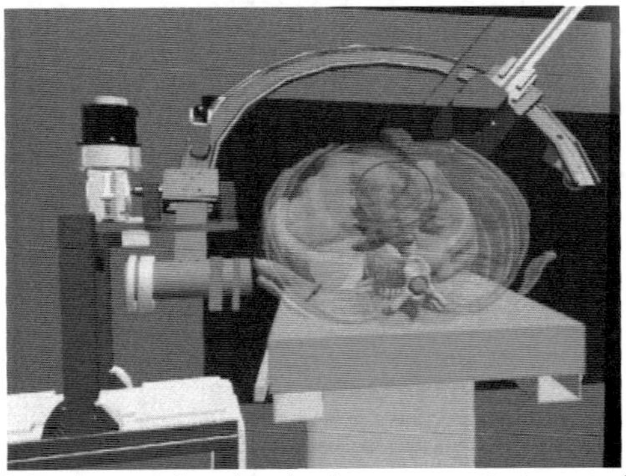

b) Task completed - the instrument is inserted in pancreas

Figure 8. a) Simulation module

3.2 Analysis of the application

The application was evaluated using a User Experience Questionnaire (UEQ) as a tool. The questionnaire was completed by 10 users in laboratory conditions (not medical procedures), and the results show that the application can be qualified as good, obtaining excellent ratings for Stimulation and Novelty (Figure 9). The main weakness identified is in terms of efficiency, which may be caused by the computational limitations of the HoloLens device [21].

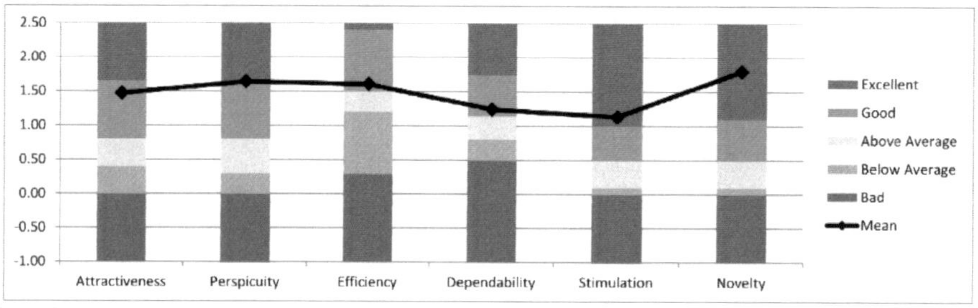

Figure 9. Results of UEQ

4. Conclusion

This paper focuses on the development and evaluation of an application, based on a mixed reality approach using Hololens 2, designed to enhance minimally invasive surgeries (MIS) by leveraging CT scanner visualization, volume rendering, and robot simulations. The goal of this application is to provide surgeons with valuable information regarding complex anatomical structures, such as the pancreas or oesophagus, thereby enhancing surgical precision and patient outcomes. However, the versatility of this application extends beyond specific anatomical regions, as it can be seamlessly integrated with various robotic platforms, allowing for its utilization across a wide range of surgical procedures. By integrating different robotic systems, surgeons can further customize their approach to each patient's unique needs, ultimately improving the overall efficiency and effectiveness of surgical interventions.

Additionally, the application provides valuable information both about the possibility of integrating robots into the operating room and about the surgery itself, enabling detailed planning. Its collaborative use between medical staff and engineers encourages beneficial cooperation that advances the development of surgical robots.

However, despite the potential benefits offered by the application, its performance is limited by computational weaknesses inherent in the device. It is crucial to consider these constraints in order to maximize the application's effectiveness in MIS procedures. Future research efforts should prioritize the optimization of computational capabilities to ensure seamless operation and user satisfaction. By addressing these challenges, our application has the potential to significantly advance surgical practice and improve patient care in the field of minimally invasive surgery.

Acknowledgment

This work was supported by the project New smart and adaptive robotics solutions for personalized minimally invasive surgery in cancer treatment - ATHENA, funded by European Union – NextGenerationEU and Romanian Government, under National Recovery and Resilience Plan for Romania, contract no. 760072/23.05.2023, code CF 116/15.11.2022, through the Romanian Ministry of Research, Innovation and Digitalization, within Component 9, investment I8.

References

[1] Attanasio, A.; Scaglioni, B.; De Momi, E.; Fiorini, P.; Valdastri, P. Autonomy in surgical robotics. Annu. Rev. Control Robot. Auton. Syst. 2021, 4, 651–679.

[2] Popescu T, Kacsó A, Pisla D, Kacsó G. Brachytherapy next generation: robotic systems. Journal of Contemporary Brachytherapy. 2015;7(6):510-514

[3] Vaida, C.; Pisla, D.; Plitea, N.; Gherman, B.; Gyurka, B.; Stancel, E.; Hesselbach, J.; Raatz, A.; Vlad, L.; Graur, F. Development of a control system for a parallel robot used in minimally invasive surgery. IFMBE Proc. 2009, 26, 171–176.

[4] Yang C, Guo S, Bao X, Xiao N, Shi L, Li Y, Jiang Y. A vascular interventional surgical robot based on surgeon's operating skills. Med Biol Eng Comput. Sep;57(9), (2019).

[5] Pisla, D.; Calin, V.; Birlescu, I.; Hajjar, N.A.; Gherman, B.; Radu, C.; Plitea, N. Risk Management for the Reliability of Robotic Assisted Treatment of Non-resectable Liver Tumors. Appl . Sci. 2020, 10, 52.

[6] Jiang, J.; Guo, Y.; Huang, Z.; Zhang, Y.; Wu, D.; Liu, Y. Adjacent surface trajectory planning of robot-assisted tooth preparation based on augmented reality. Eng. Sci. Technol. Int. J. 2022, 27, 101001.

[7] Bianchi, L.; Chessa, F.; Angiolini, A.; Cercenelli, L.; Lodi, S.; Bortolani, B.; Molinaroli, E.; Casablanca, C.; Droghetti, M.; Gaudiano, C.; et al. The use of augmented reality to guide the intraoperative frozen section during robot-assisted radical prostatectomy. Eur. Urol. 2021, 80, 480–488.

[8] Richter, F.; Zhang, Y.; Zhi, Y.; Orosco, R.K.; Yip, M.C. Augmented reality predictive displays to help mitigate the effects of delayed telesurgery. In Proceedings of the 2019 International Conference on Robotics and Automation (ICRA), Montreal, QC, Canada, 20–24 May 2019; pp. 444–450.

[9] Wörn, H.; Weede, O. Optimizing the setup configuration for manual and robotic assisted minimally invasive surgery. In Proceedings of the World Congress on Medical Physics and Biomedical Engineering, Munich, Germany, 7–12 September 2009; Vol. 25/6 Surgery, Nimimal Invasive Interventions, Endoscopy and Image Guided Therapy; Springer: Berlin/Heidelberg, Germany, 2009; pp. 55–58.

[10] Maddah, M. R., Classe, J. M., Jaffre, I., Watson, K. A., Lin, K. S., Chablat, D., Dumas, C. & Cao, C. G. (2023). A decision aid for the port placement problem in robot-assisted hysterectomy. Laparoscopic, Endoscopic and Robotic Surgery, 6(2), 43-56.

[11] Fu, J.; Palumbo, M.C.; Iovene, E.; Liu, Q.; Burzo, I.; Redaelli, A.; Ferrigno, G.; De Momi, E. Augmented Reality-Assisted Robot Learning Framework for Minimally Invasive Surgery Task. In Proceedings of the 2023 IEEE International Conference on Robotics and Automation (ICRA), London, UK, 29 May–2 June 2023; pp. 11647–11653.

[12] Hughes, B.D., Hancock, K.J., Cummins, C.B., Okereke, I.C. (2020). Surgical Approaches for the Esophagus. In: Nistor, C.E., Tsui, S., Kırali, K., Ciuche, A., Aresu, G., Kocher, G.J. (eds) Thoracic Surgery.

[13] Roulin, D., McCarron, F.N., Chirichella, T.J. et al. Robotic Pancreatic Surgery: Current Status and Techniques. Curr Surg Rep 12, 1–7 (2024).

[14] Ciocan RA, Ciocan A, Zaharie FV, Popa C, Schlanger D, Alexa A, Ionescu D, Hajjar NA. Perioperative Management of Patients with Pancreatic Cancer - the ERAS Protocols. Chirurgia (Bucur), Jun;117(4):415-422, 2022.

[15] UR5e Lightweight, versatile cobot (universal-robots.com) [accessed 18.04.2024]

[16] Saeid. H. I, H. N. D. Le, J. D. Opfermann, S. Leonard, A. Kim, M. H. Hsieh, J. U. Kang, A. Krieger, "Autonomous Laparoscopic Robotic Suturing with a Novel Actuated Suturing Tool and 3D Endoscope," in 2019 International Conference on Robotics and Automation (ICRA), , pp. 1541-1547, May 20-24, 2019

[17] Gabriela R., Iulia A., Calin V., Corina R., Bogdan G., Paul T., Alexandru P. Nadim Al H. and Doina P., High accuracy dynamically balanced spherical arm for industrial and medical applications, 2023, accepted for publications in Acta Tehnica Napocensis.

[18] Wasserthal, J., Breit, H., Meyer, M. T., Pradella, M., Hinck, D., Sauter, A. W., Heye, T., Boll, D., Cyriac, J., Yang, S., Bach, M., & Segeroth, M. (2022). TotalSegmentator: Robust segmentation of 104 anatomical structures in CT images. ArXiv.

[19] HoloLens 2— Pricing and Options | Microsoft HoloLens -https://www.microsoft.com/en-us/hololens/buy [accessed 18.04.2024]

[20] Tagkalos E, Goense L, Hoppe-Lotichius M, Ruurda JP, Babic B, Hadzijusufovic E, Kneist W, van der Sluis PC, Lang H, van Hillegersberg R, Grimminger PP. Robot-assisted minimally invasive esophagectomy (RAMIE) compared to conventional minimally invasive esophagectomy (MIE) for esophageal cancer: a propensity-matched analysis. Dis Esophagus. Apr 15;33(4), (2020).

[21] Hinderks, A., Schrepp, M., Domínguez Mayo, F. J., Escalona, M. J., & Thomaschewski, J. (2019). Developing a UX KPI based on the user experience questionnaire. Computer Standards & Interfaces, 65, 38-44.

Innovative 3D reconstruction in oral and sinus surgery using advanced printing technologies

Daria Pisla[1*], Sorana Eftimie[1*], Mihaela Hedesiu[1], Horatiu Rotaru[1], Vasile Bulbucan[2], Calin Vaida[2], Doinel Radeanu[3], Raluca Roman[1]

*Contributed equally as first authors

[1] Department of Oral and Maxillofacial Surgery and Radiology, "Iuliu Hațieganu" University of Medicine and Pharmacy, Cluj-Napoca, Romania
[2] CESTER, Research Center for Industrial Robots Simulation and Testing, Technical University of Cluj-Napoca, Romania
[3] Department of ENT Surgery, "Iuliu Hațieganu" University of Medicine and Pharmacy, Cluj-Napoca, Romania

*E-mail: dariapisla@yahoo.com, sorana.eftimie@gmail.com, mhedesiu@umfcluj.ro, roman.raluca@umfcluj.ro

Abstract. The aim of this paper is to provide an innovative approach to develop novel three-dimensional models of the paranasal sinuses in order to increase the surgical efficiency in oral and sinus surgery. The anatomic models are made from materials with special mechanical properties that mimic real human structures. The models can be used for pre-planning and training in functional endoscopic sinus surgery. The data from the CBCT (Cone beam computed tomography) scan was exported as an STL (stereolithography) file and then processed for 3D printing. A two-stage approach was used, firstly printing the sinus cavity as a rigid model to assess the geometric accuracy of the 3D reconstruction, and secondly a multi-material model with elastic properties encompassing the sinus structures as a hollow volume inside a parallelepipedal model. The preliminary tests using FESS instruments showed promising results towards the final goal to achieve an anatomic replica of the oral and sinus areas in terms of geometry and tissue elastic behaviour.

1. Introduction

Oral and sinus surgery have the role of improving the quality of life of patients with various pathologies both from a medical/functional and aesthetic point of view. The particularity and complexity of the anatomical structures in this location has determined a series of relevant research at the border of distinct fields, integrating medicine, engineering and informatics with the role of increasing diagnostic and therapeutic performance. Functional endoscopic sinus surgery (FESS) is an advanced endoscopic technique applied in the treatment of sinus defects and associated pathologies, which include congenital conditions, infectious pathology, chronic sinusitis, sinus polyps and tumors, sinus duct obstruction.

The endoscopic approach to the sinus cavities determined the decrease in the number of postoperative complications, facial aesthetic benefits (absence of scars), shorter hospitalization associated with faster recovery. However, there are aspects that can be improved [1]:

- The procedure outcome for specific individual anatomical variants determined by congenital or secondary pathologies;
- Efficient procedure preplanning to compensate for the visual operative field which is limited due to the small dimensions of the sinus structures;
- The accuracy of the procedure, considering the risk caused by the proximity of important anatomical structures at the level of the viscerocranium;
- The learning curve of this technique;
- Postoperative reconstruction.

One common solution which can contribute towards all the current limitations in sinus and oral surgery is the use of anatomic models which can be personalized (for an individual patient or a specific procedure) using different 3D printing technologies. While this approach is already used in different surgical procedures, the development of new printing technologies opens new research directions.

By creating 3D anatomical models, the advantage of using flexible endoscopes was experimentally validated, increasing the quality of intraoperative visualization. A clinical case is presented in which 3D printing technologies allowed the creation of a pre-operative model in the case of a patient with polyps that provided the necessary information related to the geometry of the tumor and subsequently allowed the establishment of an appropriate surgical protocol [3]. Another recent study, with over 800 patients, for which reconstruction at the cranial level following sinusitis was necessary, highlighted the advantage of using 3D models in the post-operative and reconstructive stage [4]. New 3D printing technologies (PolyJet) have enabled the creation of multi-material models that have improved FESS results [5]. The main 3D printing technologies existing on the market [6] are illustrated in Figure 1, highlighting the ones used in this study: Fused Deposition Modeling (FDM) or Fused Filament Manufacturing (FFF); Stereolithography (SLA); Selective Laser Sintering (SLS); Selective Laser Melting (SLM) or Direct Metal Laser Sintering (DMLS); PolyJet and MultiJet Fusion (MJF). The project will use FDM and SLA procedures for rigid structures and PolyJet technology for modeling soft mechano-elastic structures.

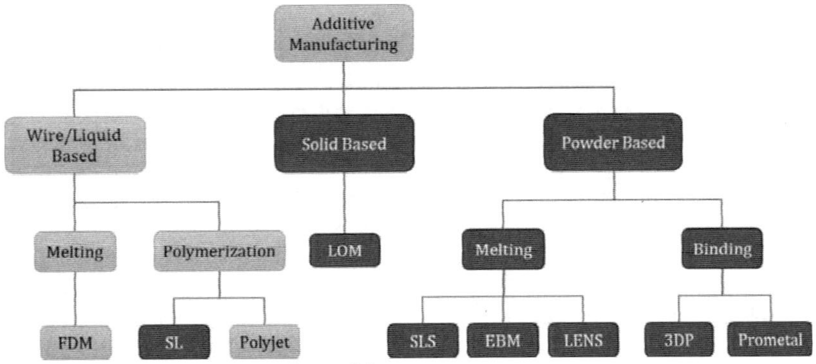

Figure 1. Additive manufacturing technologies [6]

In order to create the 3D models based on proper anatomical data, advanced imaging techniques are used to collect them and reconstruct accurate 3D models. For oral and sinus surgery, a very efficient, cost effective and patient friendly technique is the cone beam computer tomography (CBCT).

CBCT is also called volumetric dental tomographic computer and is increasingly used in diagnostics in the dento-maxillo-facial area of the body [7].

Key advantages of CBCT include a minimum radiation dose, high investigation speed, and the reduced size of the equipment, which improves the procedure ergonomics, accuracy and patient comfort. This imaging method improves the disadvantages of classical X-rays because the image is not distorted and allows the realisation of sections and reconstructions for better individualization of structures of interest. Another advantage of the CBCT is the capture of the entire volume of data required for the area of interest through a single exposure.

CBCT examinations can be of several kinds, depending on the size of the field of view (FOV). In sinus pathology, a broad FOV is required to cover the entire potentially affected area, which is a maxillofacial CBCT, where the exposure field is expanded up to the level of the Nasion anthropometric point.

Examination of the maxillary sinus by CBCT [7] is very helpful because it highlights the apex of the teeth in relation to the sinus floor and identifies the periapical lesions that may have a role in the aetiology of odontogenic maxillary sinusitis. CBCT can also determine sinus volume and lowering of the sinus floor by atrophy of the alveolar crest in patients who are fully edentulous to achieve an appropriate treatment plan.

Under normal conditions, the ciliated sinus mucosa cannot be observed radiologically. Any thickening of it over 5 mm is considered pathological and indicates a well-visible sinus inflammation by CBCT examinations.

Functional endoscopic sinus surgery (FESS) is a procedure performed under general anaesthesia, where the surgeon inserts an endoscope into the nasal cavity for sinus examination. Subsequently, various specialized instruments are used for treating various sinus pathologies or removing certain structures. While FESS has important benefits, it presents several limitations, including a steep learning curve for surgeons due to the complex anatomy of the sinus region and the proximity of vital structures (orbital walls or the skull base). Moreover, variability in patient anatomy can pose significant challenges in surgical planning and execution, increasing the risk of complications such as cerebrospinal fluid leaks and orbital injuries. Based on the current imaging technologies, like CBCT, accurate 3D digital models of the sinus area can be generated allowing then the construction of different models which can assist surgeons in the pre-planning of the FESS procedure, simulating the actual procedure of the 3D model (in complex cases) and train young surgeons to perform such surgical interventions.

In the literature, it is considered that CT, CBCT, or MRI are the most suitable imaging investigations for studying sinus pathology [1,7]. Having accurate knowledge of sinus morphology and dimensions is essential, as it can impact various therapeutic interventions. Given this anatomical diversity, a real 1:1 scale 3D model is highly beneficial to provide an accurate representation of what exists at the studied level, which offers versatility in the printing method [8,9].

In the case of CBCT sections taken at the level of the maxillary sinus, the osteo-meatal complex is targeted, where the anatomy is highly variable. (12).

The aim of this interdisciplinary research work is to construct a three-dimensional model of the paranasal sinuses based on CBCT images. This prototype is intended to serve as a guide in endoscopic surgery for personalized surgical treatment for each individual patient, aiming for the most accurate results based on the patient's individual anatomy. Additionally, the goal is to determine whether the created prototype can realistically replicate the sinus cavities from a clinical perspective.

The objective of the study is to simulate a sinus endoscopy in a 3D printed artificial environment based on the anatomical structure of the patient manufactured based on a CT or CBCT. The novelty of this work consists in development of 3D models combining two printing technologies, which allow the development of mechano-elastically model compatible with the anatomical structures, which allow large-scale studies of innovative technologies.

2. Materials and method

As input data for the 3D models, a full cranial CBCT scan was selected, where the paranasal sinuses showed no pathology, providing a reference point for individualization on pathological cases. The CBCT included in the study was obtained from a clinically healthy patient to highlight complete and structurally accurate anatomical structures. Given that it is a full cranial CBCT scan, volume segmentation was performed using the 3D Slicer program to subsequently process only the structures involving the paranasal sinuses and their adjacent areas.

The segmented data was exported as an STL (stereolithography) file, which was then printed using a 3D printer with a solid material combined with a liquid one for complex details and increased structural strength.

The aerial part of the paranasal sinuses was printed as a hard, non-elastic model, followed by the printing of its negative component, resulting in the final model. The resulting hollow structure was included in a parallelepipedal object printed simultaneously with the structures of interest, creating thus the model where the endoscopic procedure can be simulated.

The printing was done using an industrial 3D printer, Fortus 380mc, selected based on its performances in terms or accuracy and printing technique which uses two materials. From the datasheet we observe the printer is capable of layer heights ranging from 0.127 mm (T10 tip) up to 0.330 mm (T20 tip) with a tolerance of ±0.01mm. As base, an ABS (acrylonitrile butadiene styrene) material is used, and in addition, a soluble material is utilized to provide support and ensure structural integrity, with a layer thickness of 0.178 mm (T12 tip).

For the second model another printing technology had to be used, polymerization (fig. 1), which allows the controlled hardening of the deposited material. In this stage a Stratasys Polyjet J5 MediJet was used. The Polyjet technology allows the use of multi-materials in the same printing object (or assembly) allowing the construction of an object with different mechanical properties throughout its volume.

Following the 3D printing of this model, several tests were carried simulating an endoscopic procedure. The main steps of this study are presented in fig. 2.

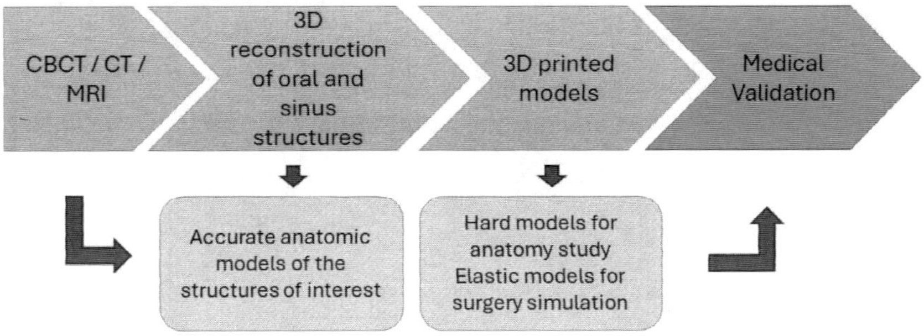

Figure 2. Design of the research study

IOP Conf. Series: Materials Science and Engineering **1320** (2024) 012011 doi:10.1088/1757-899X/1320/1/012011

3. Results

A CBCT scan from the ATLASmed database of a patient without sinus pathology (Fig. 3) was selected to be used as a reference in creating a 3D model. Volume segmentation of the paranasal sinuses was performed using the 3D Slicer program, and only the necessary data for creating the model was exported from it (Fig. 4).

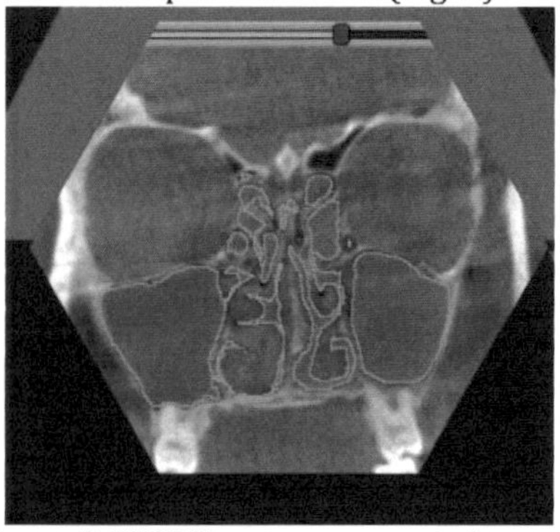

Figure 3. CBCT Structure segmentation

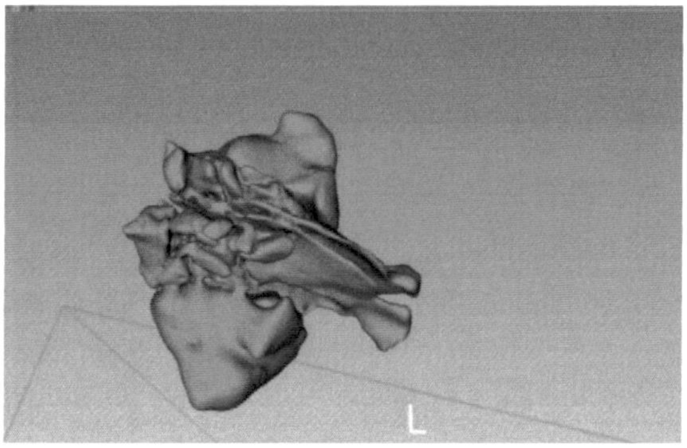

Figure 4. Volume segmentation of paranasal sinuses

The segmented data was exported as an STL (stereolithography) file, which was then printed using the industrial 3D printer, the Stratasys Fortus 380mc.

This allowed the accurate printing of the sinus cavity (as a solid model) for the geometric evaluation of the model compared to the real anatomy. The results are shown in Figures 5 and 6.

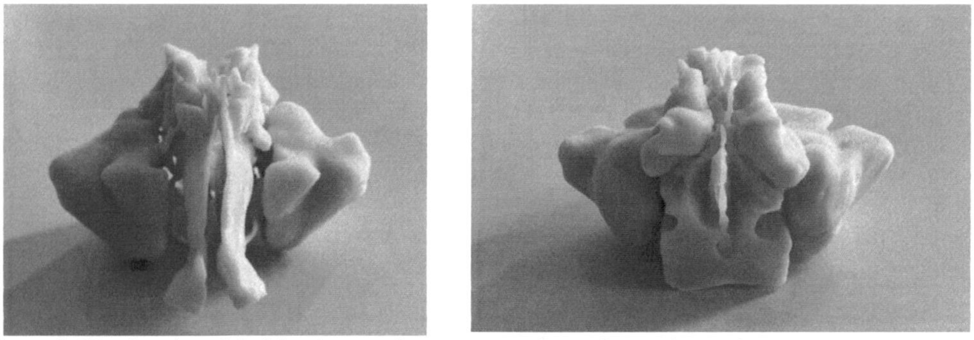

Figure 5. 3D printed model of the sinus-cavity structure – front view and rear view

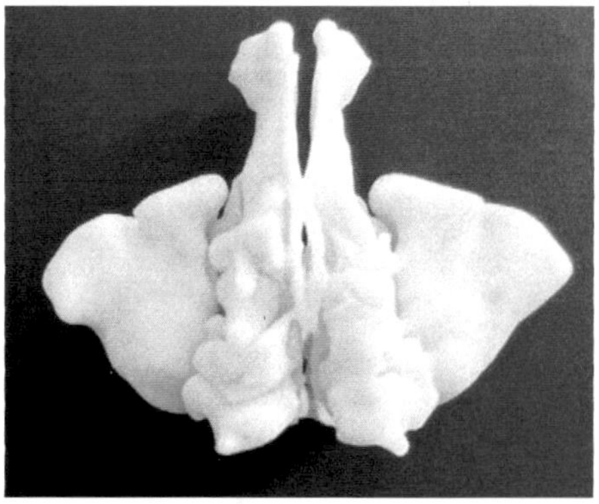

Figure 6. 3D Model of the sinus-cavity structure – top view

Once the geometry was validated, the model was used to create the negative imprint of the sinus-cavity. While the geometry was validated in this step, the challenge in printing the model with the sinus cavities referred to the achievement of the elastic behaviour of the tissues as close as possible to the real one.

The second model was printed using the Stratasys Polyjet J5 Medijet printer, which allows the controlled hardening of the deposited material. Thus, for the printing of the sinus cavity (Figure 7), the layers surrounding the hollow sinus anatomy can be printed from soft materials, a very soft (support) material is then added bordering the elastic material and all is then integrated in a harder material as support.

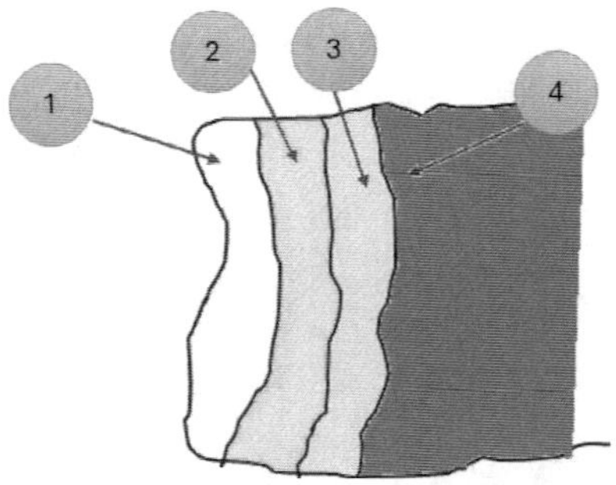

Figure 7. The layers in the 3D anatomic model (1 – hollow volume, 2 – elastic material – anatomically similar, 3 – soft, support material creating the space for the second layer to move, 4 – hard material as support)

Based on the sinus-cavity structures created on the 3D printer, a negative of these structures was made, representing the sinus cavities introduced into a parallelepiped serving as their support structure. The parallelepiped is equivalent to the framework of the human skull where the sinuses are situated (Figure 8).

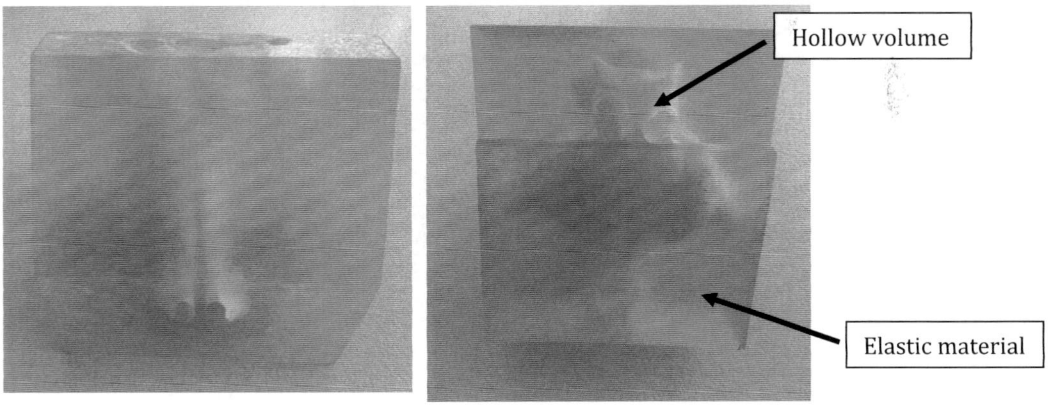

Figure 8. Front view and rear view of nasal structure

A model of the paranasal sinuses has been obtained, which exhibits a high fidelity compared to real human structures, intended for educational use. The resulted 3D printed structure was then tested in an endoscopic surgical procedure.

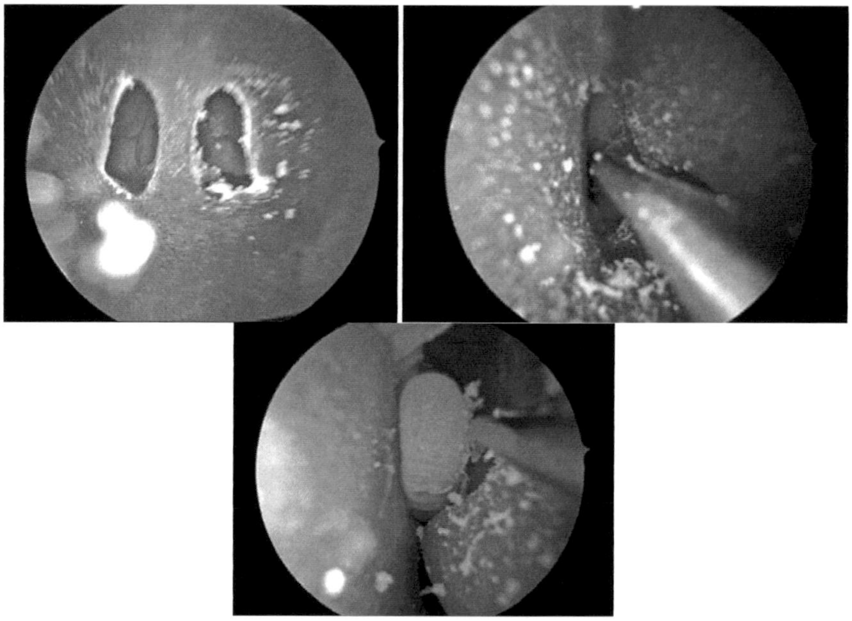

Figure 9. Endoscopic surgical procedure on developed 3D model

4. Discussions

In this study, a new three-dimensional model implemented through 3D printing is created based on previous imaging investigations. For developing the model, a CBCT of a patient that had no sinus pathologies identified was used. Following the analysis of the CBCT scan, volume segmentation was performed, retaining only the structures of interest, namely the paranasal sinuses. In the present study, this step was carried out using the 3D Slicer program. Some authors consider that the most challenging part of segmentation is represented by the upper part of the nasal cavity and the ethmoidal cells [10,11]. In the prototype created in this study, the ethmoidal cells are well represented and accurately reproduce their anatomy at the patient level, as can be observed in Figure 5.

In the literature, there are studies that have attempted to create 3D models with the aim of showing differences between individuals in sinus morphology or to study their proper and personalized irrigation [12]. In these studies, clinically healthy patients without the presence of third molars were also selected. They underwent a CBCT followed by sinus segmentation, and their paranasal sinuses were printed, with results used to compare varied anatomy. Just like in the present study, where a CBCT scan of a patient without sinus pathology was chosen, the same approach was taken in previously published studies.

Following volume segmentation, the file was exported in STL format, and three-dimensional printing was performed based on it, in a two steps approach. Segmentation represented the sinus air, and it was necessary to test its geometric quality (compared to the real anatomy) and then to create its negative to achieve a model that accurately reproduces reality. In the first printing stage, a rigid ABS material was chosen along with a soluble support, to create the 3D model representing the sinus cavity. The use of soluble support is critical for the quality of the final product, as it allows for the final processing and removal of support without generating mechanical stress, thereby preserving small details and complex surfaces.

Following the validation of the sinus geometry, a second 3D model was developed, where the sinus cavity becomes a hollow part inside the 3D part, resulting in sinus cavities and adjacent structures inserted into a parallelepiped that mimics the patient's cranial cavity, sized to encompass the interior structures of interest. In this second stage the polymerization printing

IOP Conf. Series: Materials Science and Engineering **1320** (2024) 012011 doi:10.1088/1757-899X/1320/1/012011

technology was applied, using a printer which uses multiple materials at the same time, enabling the creating of models encompassing different materials, and thus in this case, different elasticity. These materials also offer a wide range of colours to highlight different distinct anatomical parts for the most faithful reproduction of reality, and they can also be associated with different materials for the same purpose. In the 3D simulation laboratory at the University of Malaya, researchers have made various attempts at anatomical models with the aim of using real surgical instruments [13], which aligns with the purpose of the prototype developed in the current research. These models can be reusable, with the same model being used by multiple specialists in the field for educational purposes, or they can be single use, where the model is partially destroyed after simulating the intervention.

The continuous evolution of additive manufacturing technologies, such as the capability of creating multi-material models with high anatomical accuracy and the simple development of biocompatible models support the evolution of multiple medical fields towards personalized medicine. With reference to oral and sinus surgery these models can support the safer integration of FESS as a surgical technique both in terms of experts capable of performing it and its use in highly complex cases. In oral procedures, 3D printed personalised surgical guides can further enhance the patient's safety and procedure outcome. Efficient design and combination between different materials must be further explored to extend the usability of such models and create cost effective models.

One current disadvantage of creating these models lies in the cost of printing, which over time could become more accessible with beneficial implications for patients, leading to improved patient care.

The current study is still in its preliminary stages with many potential improvements, in terms of technique refining especially for the printing of the multi-material model which uses a rather new technology. However, the potential benefits of such models, as illustrated in Figure 10, support this endeavour.

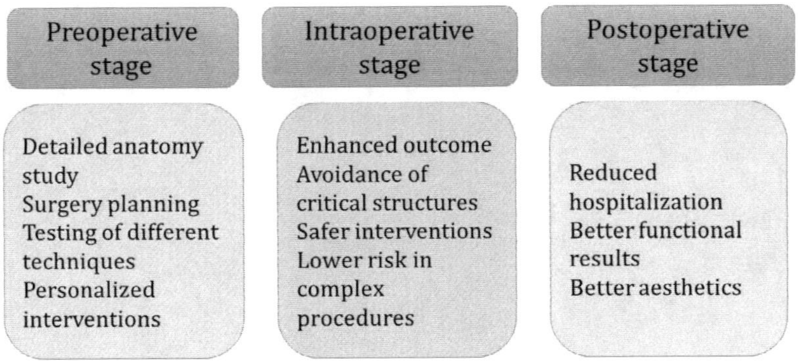

Figure 10. Potential benefits of highly accurate 3D printed models in FESS

5. Conclusion

The paper presents an innovative method to obtain a geometrically accurate 3D model of paranasal structures, providing direct visual information through accurately reproduced anatomy. Using 3D imaging, a real model can be created from different materials, offering the surgeon the possibility to better assess, plan and execute a specific procedure. Surgical skills can be transferred from the operating room to the laboratory for advance performance of surgical procedures on an accurate model, anticipating the intervention on the patient. The 3D model can also be used as a surgical guide or for educational purposes.

The preliminary tests using FESS instruments showed promising results towards the final goal to achieve an anatomic replica of the oral and sinus areas in terms of geometry and tissue elastic behaviour.

Future work targets the experimental study of the layers structuring in multi-material 3D models aiming towards a replication of the natural elasticity of the tissues and the development of anatomical molds for the study of 3D navigation techniques using custom-made modular static guides, and extended reality techniques are targeted as an innovative element in increasing precision.

Acknowledgment

This research was supported by the "Iuliu Hatieganu" University of Medicine and Pharmacy Cluj-Napoca, Romania, Department of Maxillofacial Surgery and Radiology, Doctoral Research Program, through the project PCD 646/40/11.01.2024.
This work was supported by the project Technologically Enabled Advancements in Dental Medicine (TEAM), funded by European Union – NextGenerationEU and Romanian Government, under National Recovery and Resilience Plan for Romania, contract no. 760303_CF_80_I8_R2_AA1 through the Romanian Ministry of Research, Innovation and Digitalization, within Component 9, investment I8.

References

[1] Homsi MT, Gaffey MM. Sinus Endoscopic Surgery. 2022 Sep 12. In: StatPearls [Internet]. Treasure Island (FL): StatPearls Publishing; 2023 Jan–. PMID: 33085349.

[2] Xu Z, Zhang X, Dou X, Lin C, Wang H, Song S, Yu C, Cui X, Gao X. Flexible endoscopy in the visualization of 3D-printed maxillary sinus and clinical application. Surg Endosc. 2022 Oct;36(10):7827-7838. doi: 10.1007/s00464-022-09410-8. Epub 2022 Jul 26. PMID: 35882666; PMCID: PMC9485168.

[3] Mallon A, Farnan T. A case report detailing the use of 3D printing technology in surgical planning and decision making in ENT surgery-an axial 3D first in Northern Ireland. Int J Surg Case Rep. 2021 Oct;87:106407. doi: 10.1016/j.ijscr.2021.106407. Epub 2021 Sep 15. PMID: 34560592; PMCID: PMC8473765.

[4] Gaikwad A, Malhotra R, Bikash Maiti S, Shetty AA, Rasheed DS, Kashyap L, Singh R. Role of 3D Printing in Post-op Rehabilitation of Palatal Bone Loss by Mucormycosis: A Survey. Cureus. 2022 Dec 14;14(12):e32511. doi: 10.7759/cureus.32511. PMID: 36654651; PMCID: PMC9838246.

[5] Molinari G, Emiliani N, Cercenelli L, Bortolani B, Gironi C, Fernandez IJ, Presutti L and Marcelli E (2022) Assessment of a novel patient-specific 3D printed multi-material simulator for endoscopic sinus surgery. Front. Bioeng. Biotechnol. 10:974021. doi: 10.3389/fbioe.2022.974021

[6] Rasiya, G., Shukla, A., Saran, K., Additive Manufacturing-A Review, MaterialsToday: Proceedings, 2021, 47(19): 6896-6901, https://doi.org/10.1016/j.matpr.2021.05.181

[7] Hedeşiu, M et al, Oral radiology, A practical guide for technique, anatomy and radiologic semiology, Editura Medicală publishing house, Bucureşti, 2021. (published in Romanian)

[8] Cornejo J, Cornejo-Aguilar JA, Vargas M, Helguero CG, et al. Anatomical Engineering and 3D Printing for Surgery and Medical Devices: International Review and Future Exponential Innovations. Biomed Res Int. 2022 Mar 24;2022:6797745. doi: 10.1155/2022/6797745. PMID: 35372574; PMCID: PMC8970887.

[9] Araneda N, Parra M, Valdivia Osorio J, Olate S. 3D Print of the Maxillary Sinus for Morphological Study. Int J Morphol. 2017 Sep;35(3):1102–6.

[10] Cherobin GB, Voegels RL, Gebrim EMMS, Garcia GJM. Sensitivity of nasal airflow variables computed via computational fluid dynamics to the computed tomography segmentation threshold. PLoS One. 2018 Nov 16;13(11):e0207178. doi: 10.1371/journal.pone.0207178. PMID: 30444909; PMCID: PMC6239298.

[11] Valtonen O, Ormiskangas J, Kivekäs I, Rantanen V, Dean M, Poe D, et al. Three-Dimensional Printing of the Nasal Cavities for Clinical Experiments. Sci Rep. 2020 Jan 16;10(1):502.

[12] Zhao K, Kim K, Craig JR, Palmer JN. Using 3D printed sinonasal models to visualize and optimize personalized sinonasal sinus irrigation strategies. Rhinol J. 2020 Mar 1;0(0):0–0.

[13] Complex Surgical Training with Realistic 3D Printed Simulators [Internet]. Stratasys; 2016. Available from: https://www.youtube.com/watch?v=dG2vKCkJvr4

[14] Araneda, N., Parra, M., González-Arriagada, W. A., Del Sol, M., Haidar, Z. S., & Olate, S. (2019). Morphological Analysis of the Human Maxillary Sinus Using Three-Dimensional Printing. Contemporary clinical dentistry, 10(2), 294–298. https://doi.org/10.4103/ccd.ccd_548_18

IOP Conf. Series: Materials Science and Engineering **1320** (2024) 012012 doi:10.1088/1757-899X/1320/1/012012

Wheelchair posture correction equipment

Ver Alina-Elena [1*], Ver Istvan[1] and Silviu Dan Mandru[1]

[1] Department of Mechatronics and Machine Dynamics, Technical University of Cluj-Napoca, Cluj-Napoca, Romania

*E-mail: Alina.Ver@mdm.utcluj.ro

Abstract. This paper presents an equipment for the detection and correction of the vicious posture of the wheelchair users, as well as of people who spend more time in a sitting position and need to correct their posture. Its second action is to prevent / early detect of bedsores in the sitting area. Based on the structure and components of the equipment, its functionality is emphasized and the developed prototype is presented. Some potential applications of the proposed equipment are described, for five different wheelchair vicious postures and our initial results are given.

1. Introduction

According to the World Health Organization (WHO)/2010 statistics, it was estimated that 10% of the world's population suffers from a certain form of disability, and 10% of these people need a wheelchair, that is approximately 65 million people, but according to estimates from 2023 , the number of people requiring a wheelchair has increased to 80 million, [1], [2]. At the same time, between 2012 and 2023, a decrease in the age of people affected by disability was noted: 4% increase among children, 7% increase among adults and only 2% among the older adults. [3].

1.1 From vicious posture to the risk of pathologies associated with the spine

The normal posture of the body assumes a coordinated action of locomotor apparatus and the central and peripheral nervous system, referring to the stability of the body, the balance, the relations between body segments as well as between body and the environment.

Vicious posture represents a deviation from the normal posture, which can have various causes, such as lack of or poor control over one's own attitude, repeated adoption of wrong positions and others. Healthy people who sit for a long time, or work from this position, are prone to adopting a bad posture. There are also people with congenital malformations or acquired as a result of a pathology, who adopt or are prone to adopting vicious positions.

In the case of wheelchair users, the position of the lumbar spine is most often an inclined position, which exerts additional pressure on the intervertebral discs, ligaments and fascia structures, generating and causing an imbalance of locoregional muscle tone. Shoulder pathology, due to joint overload in transfer and propulsion activities, is also not to be neglected, [4]. An equally serious problem, which is especially developed by wheelchair users, is the occurrence of bed sores in the sitting area. This occurs in pressure points developed by prolonged sitting in a vicious position in the wheelchair, the affected areas being: sacrum, coccyx and ischial tuberosities, [5].

Considering these aspects tied to the vicious posture, the equipment presented in this paper offers the possibility to correct/maintain of body posture, not only for people who use a wheelchair, but also for people suffering from various ailments at the level of the spine, such as scoliosis, hyperkyphosis, hyperlordosis, or kyphoscoliosis, or even for healthy people who want to correct their posture. Through the development of this equipment, the following objectives are achieved:

- maintaining a correct posture/as correct as possible - preventing the accentuation / erasure of the physiological curves of the spine;
- recovering and maintaining a maximum level of independence (or, depending on the pathology, maintaining this level of independence for as long as possible);
- maintaining the distance between the pubes - the xiphoid appendix;
- maintenance of respiratory volumes;
- awareness of correct posture;
- correction of vicious stations, [6], [7].

2. Constructive and functional characteristics of the equipment

The proposed equipment involves a combined action. The first function is maintaining a correct posture with the help of four ultrasonic or laser measuring sensors located in the shoulder, cervical (upper dorsal) and occipital areas, which provide colour visual feedback of the monitored areas and allows, thanks to the data recorded from the measurements, the realization of some statistics on the evolution of the posture over time. The second action is to help prevent / early detection of bedsores in the sitting area (coccyx, sacrum, tuberosities sciatica) with the help of three temperature-humidity sensors and real-time observation of the occurrence of urinary or faecal incontinence problems, which can worsen the condition of people with bedsores. The block diagram of the proposed system is given in Figure 1 and the main components are presented in Figure 2.

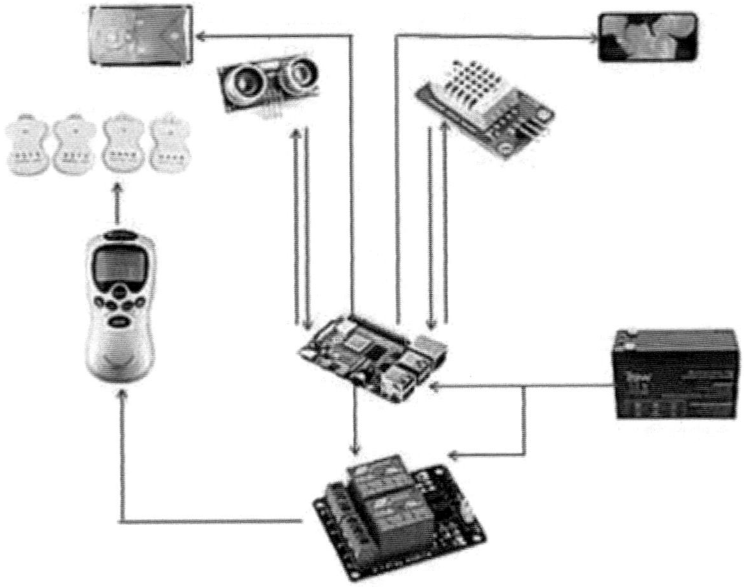

Figure 1. The block diagram of the proposed system.

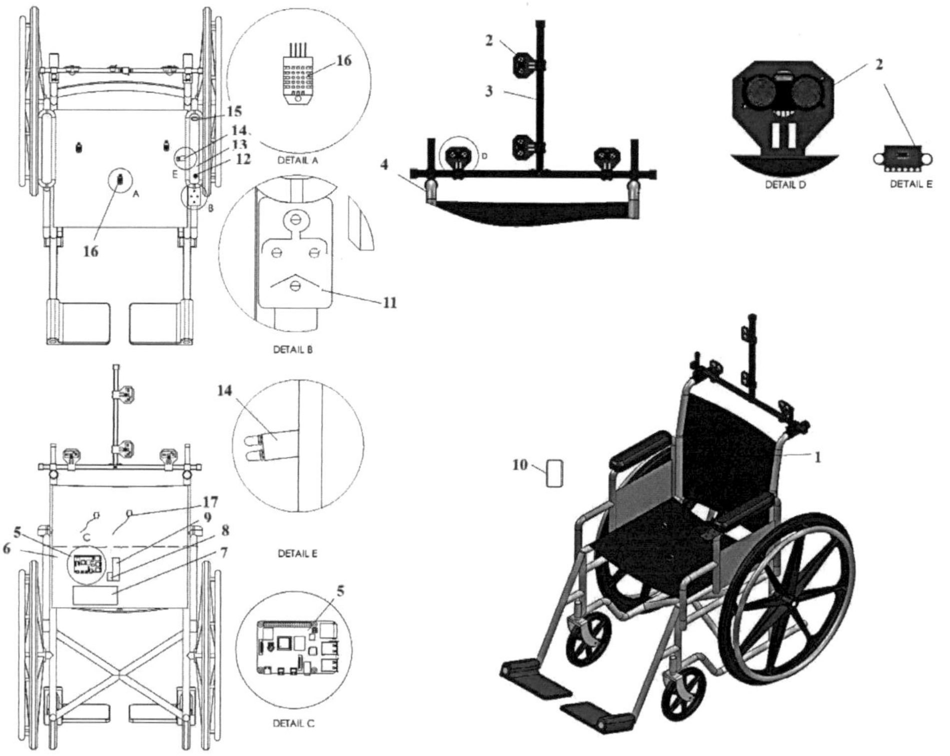

Figure 2. The equipment components: 1-wheelchair; 2-laser / ultrasonic sensors; 3- sensors support; 4-wheelchair frame; 5-development board; 6-wheelchair back pocket; 7-battery; 8-relay; 9-electrostimulation device; 10-smartphone; 11-display; 12-stop button; 13-forearm support; 14-proximity sensor; 15-electrode coupling plug; 16-temperature-humidity sensors; 17-electrodes.

The ultrasonic/laser sensors (2) are fixed with a support (3) on the frame (4) of the wheelchair backrest, allowing it to be adjusted both horizontally and vertically according to the patient's constitution. The number of ultrasonic/laser sensors is determined depending on the purpose pursued, the type and degree of the condition, etc. In the presented example, a number of 4 ultrasonic/laser sensors were used, positioned in the area of the shoulders, the upper cervical/dorsal column and the occipital region. The data taken from the ultrasonic/laser sensors (2) are processed and recorded using a development board (5) (the model chosen for example was a Raspberry Pi3b+ type), placed in the pocket (6) of the wheelchair backrest, next to the power supply (7), relay (8) and electrostimulation device (9).

Starting from the most correct position that the patient can maintain for a certain time, three measurement intervals are established, individual for the four studied sensors, presetting the distance (correct interval) and the warning time, from the smartphone (10). Based on the data measured in the set intervals, various interpretations and statistics can be made, such as: How long did the patient stay in each measurement interval during the day spent in the wheelchair? How long did it take to keep the correct posture in the set interval? The data thus collected and stored are gathered in a database, to which the specialist / patient has access through the web interface, being able to remotely follow the evolution of the posture over time. The specialist can make the most effective corrections to the therapeutic plan, without the need for the patient to go to the hospital, and the patient can follow the evolution and receive a stimulus seeing the improvement.

The feedback provided to the patient through the interface is carried out with the help of a screen (11) (or a board with LEDs - representing the anatomical shape of the monitored regions) placed on the forearm support (13) of the wheelchair, highlighting the regions and colours of the areas with problems – the LEDs used were of RGB type with 3 colours, which symbolize the areas between which the posture change is made: green-normal, blue-mediocre, red-bad. When the most vicious position is reached, after a while, the feedback is both audible (buzzer) and visual (led). Also, if necessary, upon the doctor's recommendation after exceeding the last predetermined measurement interval, if the patient does not recover voluntarily, within an initial predetermined time, after the visual and sound warning, the involuntary correction of the posture can be carried out with the help of a electrostimulation device. Thus, by connecting to a relay (8) of an electrostimulation device (9), rectangular impulses can be initialized, through which the contraction of normoinnervated skeletal muscles is stimulated - e.g. interference, which are harmless to the heart muscle and can be applied with high current intensity, causing strong, reversible, painless muscle contractions, through electrodes (17), placed paravertebrally, causing the patient to involuntarily correct his posture.

At the same time, the equipment also allows monitoring of the sitting area, with the help of three temperature-humidity sensors (16), located in the areas at risk of bedsores: sacrum, coccyx and ischial tuberosities, also providing visual feedback in the event of temperature changes being recorded . and the humidity in the respective area, on the screen (11) for the patient and at the same time, the warning will be transmitted on the smart device (10) to the staff / relatives (especially essential, in the case of patients with dementia, those in nursing homes, in hospitals).

If all monitored parameters are within normal limits, on the led interface, the patient will see all the LEDs in green, as well as the staff/relatives / caretaker as appropriate, on the web interface. To reduce the risk of recording false data and to obtain the most accurate statistics, in addition to the main switch (12), located on the forearm support, a proximity sensor (14) will be mounted under the forearm support (13), which will interrupt the operation equipment, if the patient is no longer in the wheelchair and has forgotten to disengage it.

The developed prototype is presented in Figure 3.

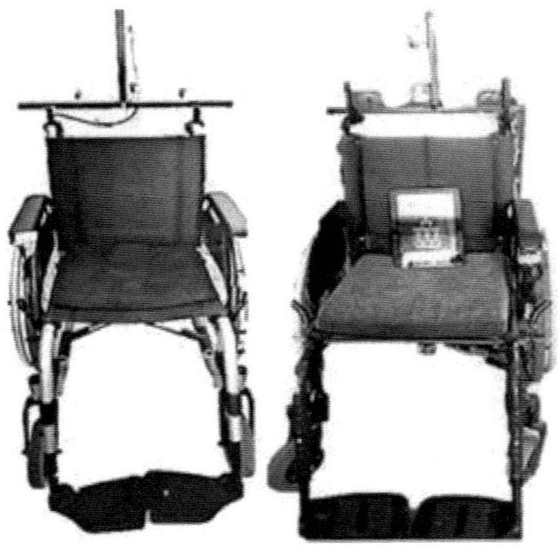

Figure 3. The developed prototype.

IOP Conf. Series: Materials Science and Engineering **1320** (2024) 012012 doi:10.1088/1757-899X/1320/1/012012

3. Wheelchair vicious postures and methods of monitoring and correcting them

Automatism is the appropriation of the nervous system (innate or acquired through exercise). Coordinated activity is an automaticity not perceived consciously but which can also be conscious, the coordination of movements being a process that can be achieved after many repetitions, [9], [10]. Thus, the subjects who use the proposed equipment by correcting their posture several times, acquire an automaticity in maintaining a correct/as correct as possible posture. Table 1 gives some potential applications of the proposed equipment.

Table 1. Applications of the proposed equipment.

Deviation type	Image	Characteristics	Monitoring
A. Posterior pelvic tilt with thoracic kyphosis and forward flexion of the neck		- excessive thoracic kyphosis, leads to a "C"-shaped spine, to "flattening" the lordosis of the cervical and lumbar spine; at the level of the head, the decrease in cervical spine lordosis causes the neck to flex forward and look down towards the floor	- forward flexion of the neck will be monitored, through 2 sensors: at the level of the lower cervical / upper dorsal spine
B. Anterior pelvic tilt with lumbar lordosis and neck hyperextension		- excessive lordosis of the lumbar spine and cervical spine; - excessive lordosis of the cervical spine, causes hyperextension of the neck and looking up	- monitoring will be done through all 4 sensors
C. Pelvic rotation with spinal rotation and lateral neck flexion		- the thoracic spine follows and rotates in the same direction as the pelvis	- monitoring will be done through all 4 sensors
D. Pelvic obliquity with scoliosis and lateral flexion of the neck		- when one side of the pelvis is raised higher than the other, the thoracic spine curves creating a scoliosis; the lateral flexion will usually be towards the side where the hip is higher	- monitoring will be done through all 4 sensors
E. Compound deviation - usually presents itself as a rotation or scoliosis, depending on the position of the pelvis.	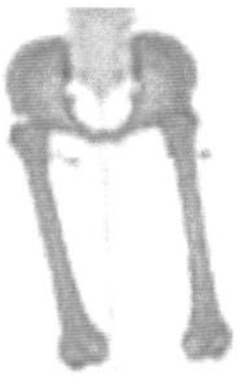	- the neck will enter lateral flexion, a position that can be accompanied by forward flexion of the head	- monitoring will be done through all 4 sensors

In cases B, C, D and E, an important role is played by the sensor in the occipital region which, when adopting the position with the neck in lateral flexion, will detect the exit from the surveillance area, as well as the sensors in the shoulders, which will also notice these exits, [11] - [16]. For the accuracy of the records, tests were carried out with two types of measuring sensors – ultrasonic sensors and laser sensors, on various materials: cotton, silk, synthetic material and cotton-synthetic material combinations. The test results, as well as the technical characteristics of the two types of sensors, showed that laser sensors provide more accurate results than ultrasonic sensors.

The equipment allows easy setting of the parameters of the measuring sensors (measuring range, warning time) as well as the temperature-humidity sensors. All settings can be made remotely, in real time, from a smart device (smartphone / tablet / laptop). Due to the possibility of recording, storing and processing the collected data, various statistical interpretations of the evolution of the patient's posture over time can be made.

The web interface and distance setting panel are given in Figure 4, and some results are shown in Figures 5 and 6.

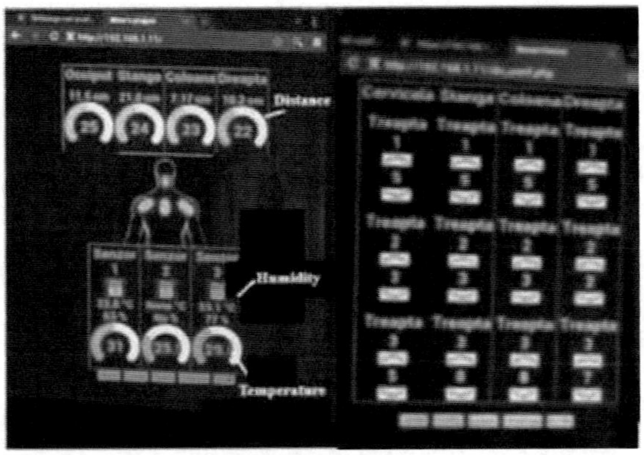

Figure 4. The interface.

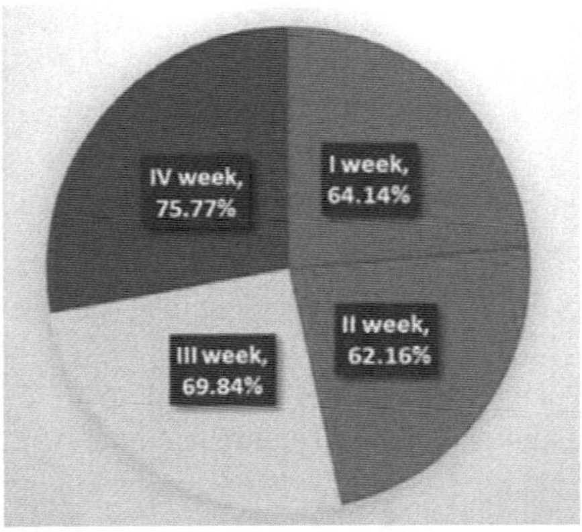

Figure 5. The average percentage of time spent in the correct posture

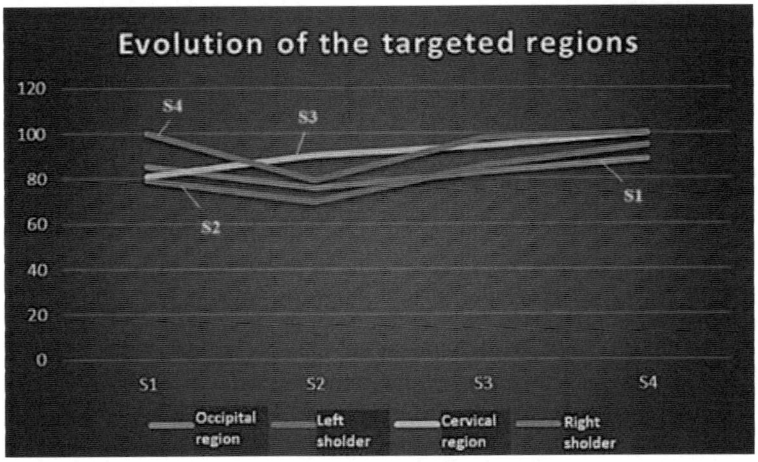

Figure 6. The evolution of targeted regions in the correct position during the period.

4. Conclusions

The developed equipment can be used by a a large number of patients, with various pathologies at various stages, but it can also be used by people who spend more time in a sitting position, and intend to correct their posture.

The main functions of the proposed therapeutic equipment are:

- recording and transmitting the collected data to the medical staff, in order to interpret them ;
- remote modification of the parameters of the recovery program, depending on the evolutionary stage or necessity ;
- the possibility of real-time assessment of the patient's posture ;
- the equipment allows the therapy to be carried out in a family setting, contributing to the efficiency of recovery .

Due to the possibility of using the equipment at home with minimal training of the patient, the pressure on the medical personnel outnumbered by the number of patients is also reduced. The doctor/therapist can monitor several patients remotely, having access to real-time data and being able to modify the parameters of the therapeutic program remotely in real time, the only restriction being given by the patient's/therapist's access to the Internet. Moving patients in a wheelchair is difficult (for example, electric wheelchairs require special means of transport, because there is no possibility of folding them), moving with the help of public transport also increases stress and can be an impediment to the continuation of therapy.

Other functional characteristics are:

- the device shows easy adaptability of the measuring sensors to any type of chair, office armchair or wheelchairs and has battery autonomy;
- early detection of bedsores, prevents aggravation of conditions

it helps to avoid these patients becoming immobilized in bed, unable to use the wheelchair, thus reducing their mobility and independence, or even worse, death;

- thanks to the proximity sensor, posture data is collected only while the patient is in a wheelchair.

References

[1] https://iris.who.int/bitstream/handle/10665/205041/B4616.pdf?seq Apr 26, 2024
[2] https://www.who.int/teams/health-product-policy-and-standards/assistive-and-medical-technology/assistive-technology/wheelchair-services Apr 26, 2024
[3] https://www.activityalliance.org.uk/news/8871-number-of-disabled-people-in-uk-increasing-over-time 21.03.2024 pdf Apr 26, 2024
[4] Ungur R, Maniu D A, Ver A, Irsay L, Ciortea V and Borda I M 2018 The major risks associated with manual wheelchair use Approaches and methods in the habilitation and rehabilitation of people with disabilities editors Maniu A E and Maniu D A (Cluj-Napoca: Risoprint) p 241-243
[5] https://fefsoradea.ro/fisiere/cadre/5_Kt_def_fizice_Sistemul_postural.pdf Apr 26, 2024
[6] https://reabilitare.usmf.md/wp-content/blogs.dir/129/files/sites_/129/2018/04/Locomotor.pdf 2024
[7] Haas B, Playford ED, Ahmad AQ, Yildiran T, Gibbon AJ, Freeman JA. 2016 Rehabilitation goals of people with spinal cord injuries can be classified against the International Classification of Functioning, Disability and Health Core Set for spinal cord injuries. Spinal Cord. 2016;54(4):324-328. doi:10.1038/sc.2015.155
[8] Ciortea VM 2017 Electrotherapy (Cluj-Napoca: Medical University Iuliu Haţieganu Publishing House)
[9] https://dexonline.ro/definitie/automatism/902605 Apr 26, 2024
[10] Sbenghe T 1987 Therapeutic and rehabilitative practical kinesiology (Bucharest: Medical Publishing House)
[11] https://hub.permobil.com/blog/common-postural-abnormalities-wheelchair-seating-positioning 2024
[12] Sprigle S, Wootten M, Sawacha Z and Theilman G 2016 Relationships Among Cushion Type, Backrest Height,

Seated Posture, And Reach Of Wheelchair Users With Spinal Cord Injury, pp 236-243 https://doi.org/10.1080/10790268.2003.11753690,

[13] http://www.sunrisemedical.ca/dealer-clinician-tools/education-in-motion/clinical-corner/march-2011/practical-seating-considerations-pelvic-obliquity Apr 26, 2024

[14] https://centrokinetic.ro/fiziokinetoterapia-explicata/totul-despre-recuperare/cifoza-cauze-tratament Apr 26, 2024

[15] Thomson D, Tully P and Blochlinger S 2013 Laying the foundation for proper positioning: Introduction to positioning for functional ability and wheelchair seating https://www.seatingsymposium.com/images/pdf/PS/PS2_Tully_Thomson_Blochlinger.pdf Apr 26, 2024

[16] Yang YS, Chang GL, Hsu MJ and Chang JJ 2009 Remote monitoring of sitting behaviors for community-dwelling manual wheelchair users with spinal cord injury. Spinal Cord. 2009;47(1):67-71. doi:10.1038/sc.2008.72

IOP Conf. Series: Materials Science and Engineering **1320** (2024) 012013 doi:10.1088/1757-899X/1320/1/012013

Chemical biosensor for detecting immunity against viral diseases (SARS-COV2)

Anitei Vlad-Alexandru and Mihai Munteanu

1 Faculty of electrical engineering, TECHNICAL UNIVERSITY OF CLUJ-NAPOCA, Romania

*E-mail: vlad.anitei@umfst.ro

Abstract. Due to the need of developing means of rapid testing for SARS COV 2 virus infection, a series of methods for fast investigation of individuals have been created. In this regard, this article proposes a colorimetric testing method for the infection, using a chemically modified paper support and an RGB analysis method to identify positive tests.

1. Introduction

In the wake of the unprecedented COVID-19 virus pandemic, global efforts to combat the virus have led to remarkable innovations across various scientific disciplines. One of the most notable advances in the field of diagnostics has been the development of paper-based COVID-19 tests, which have become a common solution for rapid and accessible detection of the SARS-CoV-2 virus. These tests offer a combination of accessibility, simplicity, and efficiency, addressing some of the main challenges associated with widespread testing.

Similar to other coronaviruses, the one that triggered the 2020 pandemic includes several crucial proteins, among them a spike protein (S), a glycoprotein membrane protein (M), an envelope protein (E), and a nucleocapsid protein (N). These proteins play vital roles in facilitating infection. For example, the spike protein allows the virus to enter host cells by binding to the host's angiotensin-converting enzyme 2 receptor (ACE2) and facilitates the fusion of viral and host membranes. [1]

Cellulose paper has been used as a substrate for sensory applications since the 19th century, with urine and pH testing strips being excellent examples of its use in this regard. Its importance lies in its role as a support material for separating components from complex matrices, particularly in techniques such as chromatography and electrophoresis.[2] Cellulose possesses remarkable attributes that align with the principles of green chemistry, including regenerability, recyclability, and biodegradability. It is also one of the most abundant biopolymers on Earth.

During the COVID-19 pandemic, several testing methods were developed and widely used for detecting and monitoring the spread of the virus. The "gold standard" of these tests was the RT-PCR (Reverse Transcription Polymerase Chain Reaction) test, which involves collecting a sample from a person's nose or throat using a swab, which is then processed in a laboratory to detect viral RNA. [3]

It is also worth mentioning antigen tests, which detect specific viral proteins and are faster than PCR tests, providing results in minutes. They are less sensitive than PCR tests but are useful for rapid screening in various settings.[4]

Thus, the research work consisted of developing a testing strip on a cellulose (chemically treated paper) substrate for rapid and quantitative detection of antibodies against SARS-CoV-2 by visually comparing color changes. For this purpose, cellulose paper is chemically modified by treatment with sodium metaperiodate (NaIO4) to introduce aldehyde groups for additional covalent bonding of biorecognition elements.

Color detection was achieved with a conventional staining solution (Ponceau S solution). Color changes were recorded with a smartphone camera, with the intensity of the color indicating an increase in antibody concentration. Consequently, a successful relationship can be established between antibody concentrations and color intensity (RGB values of the captured images). The high precision of this method has been demonstrated to be effective (linear response of RGB values) for serum samples with concentrations from 10 ng/mL to 1 μg/mL.

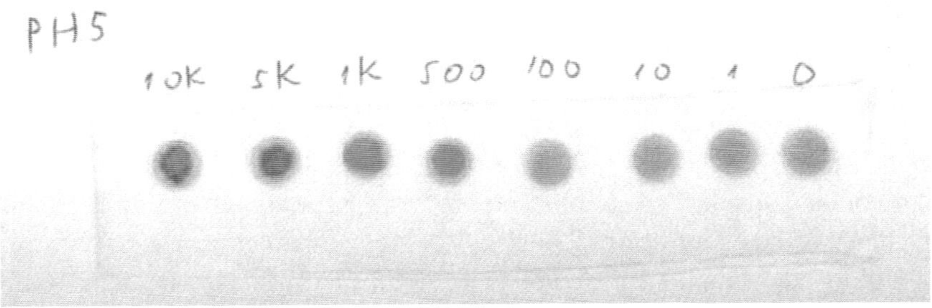

Figure 1. Example of a successful colorimetric scale.

2. Equipment and technology

2.1 Bradford coloring agent

The Bradford method is a common technique used to quantitatively determine the concentration of protein in a solution. It relies on the interaction between proteins and a dye called Coomassie Brilliant Blue, which results in a color change that can be measured spectrophotometrically. Coomassie Brilliant Blue exists in a red form (absorbing at 465 nm) and a blue form (absorbing at 595 nm). When the dye binds to proteins, it shifts from the red to the blue form, resulting in increased absorption at 595 nm. These values indicate the wavelengths of light that the dye absorbs when it binds to proteins, changing its color and thus allowing for the quantitative measurement of proteins. [5]

The Bradford reagent, often called Bradford reagent or Coomassie dye, is a solution containing Coomassie Brilliant Blue and other necessary chemicals for the Bradford method. The dye interacts with protein molecules, resulting in a color change that is directly proportional to the protein concentration. This color change can be quantified using a spectrophotometer, allowing researchers to determine the protein concentration in a sample.

IOP Conf. Series: Materials Science and Engineering **1320** (2024) 012013 doi:10.1088/1757-899X/1320/1/012013

2.2 Elements used in data interpretation

For the human protein test, tests were conducted in pH environments of 5 and 7, related to the isoelectric point of antibodies. The isoelectric point is a fundamental concept in biochemistry referring to the pH at which a molecule carries no net electric charge. It is a critical property of molecules containing acidic and basic functional groups, such as amino acids, peptides, and proteins. At the isoelectric point, the molecule is electrically neutral because the number of positive charges (protons) equals the number of negative charges (electrons), resulting in a net charge of zero[6]. Since the isoelectric point of antibodies is around 5, below this value, antibodies will have a positive charge, and since the dye is highly negative, the interaction between the two will theoretically be stronger.

For the majority of the human protein tests, spike protein (S protein) was used. The spike protein is a surface protein of the SARS-CoV-2 virus, responsible for the COVID-19 pandemic. This protein is essential for the infection of human cells by the virus as it binds to receptors on the surface of human host cells, allowing the virus to enter the cell and multiply. Due to its key role in infection, the spike protein has been a major target for vaccine development and other therapies during the pandemic. [7]

The used concentration for the experiments was 5μL/mL (meaning 5 μL of protein serum solution per 1 mL of basic solution). Since all protein samples in the laboratory stock have a concentration of 50 μL/mL and are divided into amounts of 20μL, 180μL of buffer solution was added to achieve the desired concentration. 15μL of the resulting solution was applied to each paper and placed in the refrigerator for 24 hours to allow the protein incubation process.

For a more accurate visualization of the color scale, a broader range of concentrations was used (50,000μg/mL, 10,000μg/mL, 5,000μg/mL, 500μg/mL, 100μg/mL, 50 μg/mL, 10 μg/mL, 5 μg/mL, and 0 μg/mL).

The block diagram of the implemented application, based on which the entire experimental process of detecting SARS-CoV-2 infection is conducted, is presented in Figure 2:

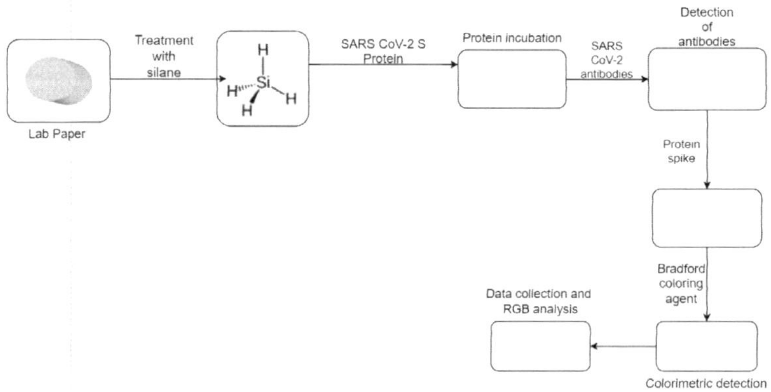

Figure 2. Diagram of the entire experimental process.

The paper is treated with an aminopropyltriethoxysilane solution, after which human protein is applied and left to incubate for at least 24h in the refrigerator. The samples are then treated with Cov-2 antibodies in various concentrations which are ultimately illustrated by the coloring agent. The colorimetric scale of the samples is photographed and digitally analyzed.

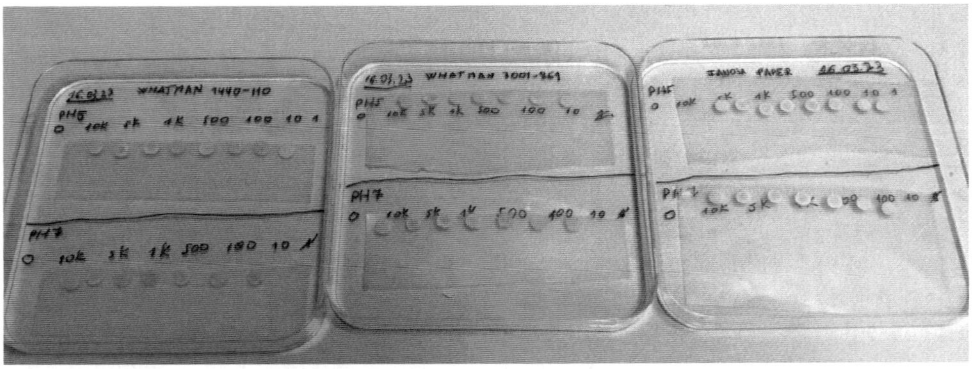

Figure 3. Picture of 6 different batches of tests

In the above presented picture there a set of 6 tests is exemplified, with various types of paper and PH levels for the basic solution used. The picture was taken after finishing the drying process in a controlled temperature oven.

2.3 Data collection software components

Regarding the analysis and data collection to automate the process as much as possible, a system was designed to capture images of paper tests using an OV7670 camera controlled by an Arduino UNO board. The photos stored on the board's internal memory are sent via an Arduino script to a local file through a USB connection. Eventually, a Python program retrieves the images, analyzes the RGB spectrum of the image pixel by pixel using a "pixel_value[x][y]" function, and writes the average for each primary color into an Excel file.

To ensure the quality of the analysis, the images were also analyzed using the open-source software 'ImageJ'.

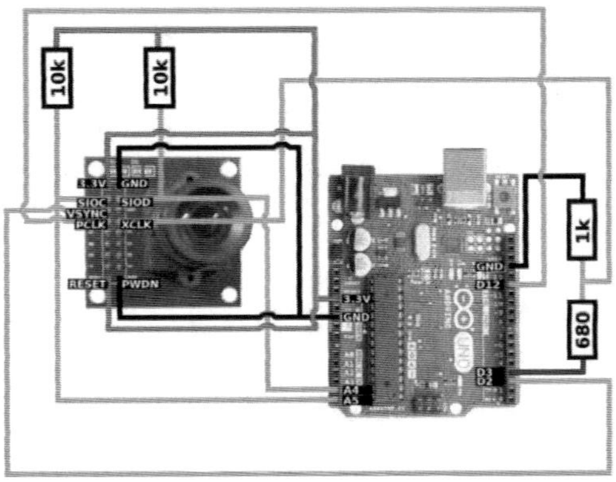

Figure 4. Arduino wiring diagram

3. Interpretation of the results

The level of success, in an experiment or the extent of linearity is indicated by R^2. R^2, also known as the correlation coefficient is a metric that illustrates the percentage of variability in a variable that can be accounted for by an independent variable or variables in a regression model. Essentially R^2 offers insight into how the data align with the projected model (with 0 indicating that the model does not clarify any variations in the variable and 1 indicating that the model fully elucidates the variations, in the dependent variable).A test is considered successful when the correlation coefficient is greater than 0.9 and ideally exceeds 0.98. Working with multiple data sets has shown that the best variable for graphical representation is the average green value divided by the average blue value. Naturally, the more visibly the colorimetric scale is to the naked eye, the closer, in theory, the correlation coefficient tends to 1.

For an experiment to be considered scientifically relevant, its reproducibility must be demonstrated. For this test with human protein, paper treated with silane and enriched with 5μg/mL was used. The data comes from pictures taken 30 minutes after the application of the coloring agent. The variable considered to most advantage the results is obtained by dividing the blue value by the green for each concentration by the blue divided by green for the control sample (concentration 0).

To better represent linearity, outlier values, which deviated too much from the scale, were removed from the graphs. For the final result to be considered valid, considering a test has a number of 9 colorimetric values, a maximum of 3 values can be removed. Eliminating more values would make obtaining linearity irrelevant, losing too significant a percentage of data.

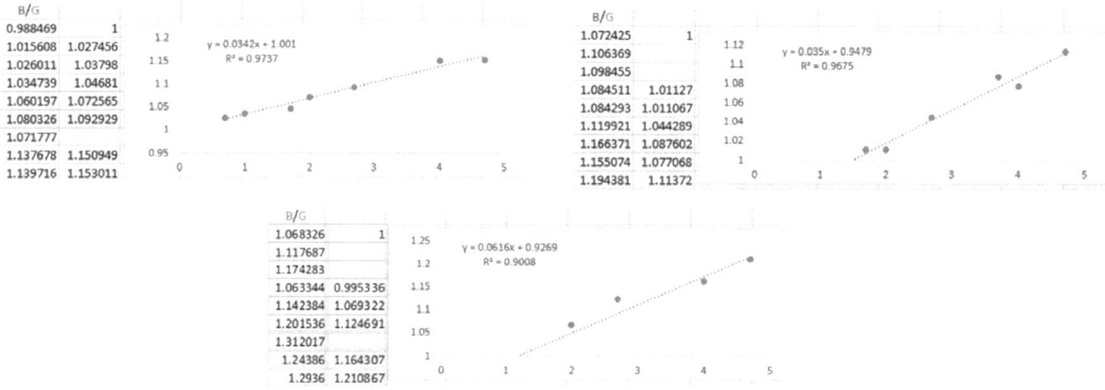

Figure 5. The obtained values and their graphical representation for the three iterations of the mentioned test.

Each graph shows the colorimetric scale (the ratio between the RGB blue value and the green value) for samples of increasing concentration (from 0 to 50,000 μg/mL). For each experiment, a number of values were removed from each graph to achieve the best linearity possible. As a result, correlation coefficients of 0.97 (top left graph), 0.96 (top right graph), and 0.9 (bottom graph) were obtained.

To obtain a visualization, a graph was created that represents the average values for each concentration of the three tests previously presented. Above the graph, the colorimetric scale can be observed, which becomes noticeably darker as the concentration of the solution increases.

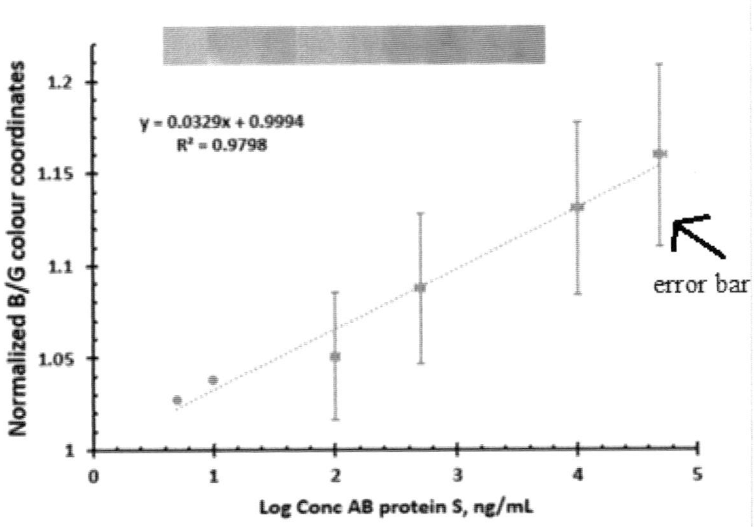

Figure 6. Graphical representation of the average of the most successful series of tests, with a correlation coefficient of 0.9798

The chart above shows a correlation coefficient of 0.9798 indicating that, around 97.98% of the variation in the normalized / color coordinates can be explained by the logarithm of protein S concentration. This demonstrates a correlation and high predictability of color coordinates based on the independent variable.

The equation y=0.0329x+0.9994 represents how the logarithm of protein S concentration relates to the normalized / color coordinates. The steeper slope (0.0329) compared to the graph suggests a significant increase in color coordinates for each unit rise in protein S concentrations logarithm.

Additionally error bars are used to plot normalized B/G color coordinates, against AB protein S concentration (ng/mL). These bars show the expected range within which each data points true value's likely to lie.

4. Conclusions

The two elements that most helped achieve consistent positive results are temperature control and pre-enrichment of the papers before applying different concentrations. Through a process of 'trial and error,' it was determined that a concentration of 5µg/mL for the enrichment solution is ideal for making the colorimetric scale more visible, while a concentration of 1µg/mL was insignificant and one of 10µg/mL disturbed the color formation too much, especially for lower concentrations. Regarding temperature control, this was the essential factor in increasing the success rate of the tests. Color formation for the tests with protein S is much more sensitive compared to tests using animal protein. Therefore, a laboratory oven was used at a constant temperature of 15°C. This increased the color development time by about 20 minutes, which is detrimental to the research: developing a rapid-action paper-based Covid test. A possible solution could be experimenting with different constant temperatures at which the paper is dried.

The same testing methodology can be applied for the N protein of the virus. The nucleocapsid protein, often abbreviated as the N protein, is another essential protein of the SARS-CoV-2 virus. It plays a role in the viral life cycle, particularly in packaging viral RNA (genetic material) into new viral particles. The nucleocapsid protein is located inside the viral envelope and is not exposed on the surface of the virus, as is the spike protein (S protein). It is not a primary target for vaccines, as most vaccines aim to stimulate an immune response against the spike protein, which is the main mode of entry into human cells. However, COVID-19 tests often detect the presence of the nucleocapsid protein or viral RNA as part of the diagnostic process [8].

The testing method presented in the article offers two major advantages. Firstly, unlike the classic test, which provides a binary result (positive or negative), the tests presented offer a perspective on the level of infection in the body (concentration level). Secondly, having a cellulose-based paper, the test is very cheap to produce and apply. Regarding the practicability of the test, more work needs to be done to increase accuracy and reduce the rate of failed tests.

As a future direction, the research can be advanced with an FTIR (Fourier Transform Infrared Spectroscopy) analysis of the samples after they have completely dried, in order to identify the types of chemical bonds and functional groups in the molecules of the test paper.

References

[1] Protein & Cell, volume 11, issue 4, page 235, April 2020, "Puzzle of highly pathogenic human coronaviruses (2019-nCoV)"

[2] Microchimica Acta, volume 187, article number 272, 2020, "Monitoring cellulose oxidation for protein immobilization in paper-based low-cost biosensors"

[3] The Protein Protocols Handbook, pages 17-24, 2009, "The Bradford Method For Protein Quantitation"

[4] M. Smoluch, 2016, "Proteomic Profiling and Analytical Chemistry"

[5] Annual Review of Biochemistry, Volume 45, pages 217-238, 1976, "Comparative Aspects of Glycoprotein Structure"

[6] Journal of Microbiology, Immunology and Infection, Volume 54, Issue 2, 2021, Pages 159-163, Leila Mousavizadeh, Aprilie 2021, "Genotype and phenotype of COVID-19: Their roles in pathogenesis"

[7] Eurosurveillance, volume 25, issue 3 Corman VM, 2020, "Detection of 2019 novel coronavirus (2019-nCoV) by real-time RT-PCR."

[8] Cochrane Database of Systematic Reviews, Issue 3, article number CD013705, 2021 Dinnes J, "Rapid, point-of-care antigen and molecular-based tests for diagnosis of SARS-CoV-2 infection."

IOP Conf. Series: Materials Science and Engineering **1320** (2024) 012014 doi:10.1088/1757-899X/1320/1/012014

Impact of a MATLAB-controlled application on the autonomy of ischemic stroke patients in the hospital

D. A. Ivascu[1] and M. Munteanu[2]

[1,2] Technical University of Cluj-Napoca, Faculty of Electrical Engineering, Romania

*E-mail: danieladrian.ivascu@yahoo.com

Abstract. This study explores the possibility of using a Matlab-controlled electromyographic system, based on a microcontroller, in monitoring and assisting patients with contralateral hemiplegia following an ischemic stroke, paralyzed on one side of the body and bedridden. The main aim is, by generating electrical signals based on the level of flexion of the patient's healthy arm or clenching of the patient's healthy fist, the system can control various devices and alert medical staff. The technology provides support and independence to patients in performing some of their daily activities.

1. Introduction

Ischemic stroke is one of the most common and devastating neurological conditions globally, affecting millions of people each year. According to recent data, ischemic stroke accounts for approximately 80% of all stroke cases and is one of the leading causes of disability and mortality worldwide [1]. This condition occurs as a result of obstruction or reduction of blood flow to a specific part of the brain, leading to oxygen and nutrient deprivation in that area [2]. The consequences can be extremely severe, including motor deficits, speech disorders, cognitive impairment, and the need for long-term care [3].

In addition to the direct impact on the health and quality of life of patients, ischemic stroke also has significant consequences on society and the economy. The costs associated with the treatment and care of ischemic stroke patients are considerable, and the productivity losses caused by long-term disability can have a substantial impact on the economy [4]. In this context, research aimed at improving the diagnosis, treatment and recovery of ischemic stroke patients, this being a matter of crucial importance.

Thus, the use of assistive technologies can play a vital role in improving the quality of life for ischemic stroke patients. Assistive technologies, such as Arduino-controlled electromyographic devices, can provide patients with support and independence in performing daily activities and reintegrating into the community. These technologies can facilitate therapeutic exercises and real-time progress monitoring, thus promoting the recovery and rehabilitation of patients.

Moreover, assistive technologies can contribute to reducing patients' dependency on caregivers and constant assistance, allowing them to maintain as much autonomy and independence as possible in their daily lives [11].

2. General considerations regarding electromyography

As a method, electromyography is widely used for assessing the electrical activity of muscles. Its main applicability lies in the medical field, particularly in the diagnosis and monitoring of neuromuscular disorders, as well as in the physical rehabilitation of patients with various conditions, including ischemic stroke.

When a muscle contracts, its muscle fibers generate electrical impulses called action potentials. These electrical signals can be detected by electrodes placed invasively or on the external surface of the body over the muscles under investigation (surface electromyography, sEMG) and the recorded signals are then processed and interpreted to provide information about muscle activity [5], [6].

The EMG signals detected by electrodes are generally of low amplitude and can be affected by electrical noise and external interferences. Therefore, the signals are amplified and filtered to improve their quality and clarity[8]. EMG processing may involve the use of signal analysis algorithms to identify specific patterns and characteristics of EMG signals, as well as to quantify parameters such as frequency, amplitude, and duration of muscle contractions.

An approach to evaluate electromyography can be realised the use of temporal integration of the signal, which measures its integrated value on a specific time interval, a technique that provides a clear indication of muscle activity, but that can be affected by electrode placement[9].

In the context of ischemic stroke, electromyography can be used to assess and monitor the motor function affected by patients. EMG recordings can provide information about the degree of paralysis or muscle weakness and can be used to guide the rehabilitation plan and physical therapy of patients [10].

3. The purpose and relevance of the proposed study

In light of these considerations, the presented study aims to explore the effectiveness of using a Matlab-controlled electromyographic system based on Arduino in monitoring and assisting ischemic stroke patients in an individual hospital setting. By generating electrical signals based on the flexion level of the patient's unaffected arm, the study will investigate whether this system has the potential to provide support and independence in daily activities and thus promote the recovery and rehabilitation of patients during hospitalization. Thus, by clenching the fist of the unaffected arm, the patient will be able to alert medical staff when needed, and by flexing the same arm, the patient will be able to activate a ventilator or air conditioner, which will be turned off when temperature and humidity values reach the set reference points.

3.1 General Objectives

Precise monitoring of muscle activity is one of the evaluation criteria for this application, alongside feasibility and adaptability in a real clinical setting. These include cost analysis, implementation time, staff training, and identifying obstacles or potential difficulties in use and integration into daily routine, all leading to assessing the level of independence and comfort of patients in performing daily activities.

3.2 Specific Objectives

Investigating the operation of the proposed application in detecting the level of arm flexion and transmitting signals to the Atmega 328 microcontroller focuses on the process of capturing electrical signals generated by muscle activity and converting them into interpretable digital signals for initiating appropriate actions.

To ensure an appropriate response to variations in electrical signals produced by muscle contractions, sensor calibration is necessary.

Validation of this process is achieved through controlled experiments, monitoring electromyographic signals while patients perform specific arm movements. These experiments may involve both healthy subjects and ischemic stroke patients to assess the accuracy and sensitivity of the proposed application in detecting muscle activity.

4. Method Description

The proposed system for assisting patients with ischemic stroke utilizes a combination of dedicated hardware and software solutions to improve the quality of care and patient comfort, and the implementation steps are outlined below.

4.1 Components

The main components and functionalities of the system include:

- The Arduino platform as the Central Control System, includes the Atmega microcontroller, responsible for coordinating all functions and devices within the application. It receives and processes signals from the electromyographic sensors and the environmental sensor, making real-time decisions and controlling external devices based on the received data.
- Electromyographic Sensors for Monitoring Muscle Activity: The signal electrodes are placed on the surface of the patient's arm to monitor muscle activity. The electrical signals generated by muscle contractions are captured and processed to detect and quantify the level of arm flexion.

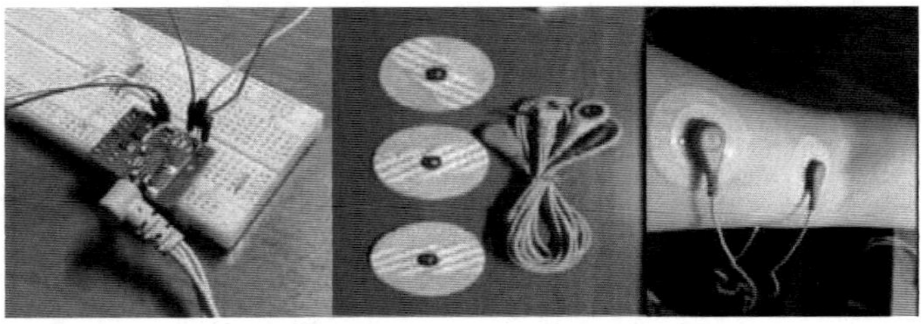

Figure 1. EMG sensor, surface electrodes, arm positioning of electrodes.

- Ambient Sensor DHT22 for Environment Monitoring that constantly monitors the temperature and humidity in the room. This data is used to ensure a comfortable and healthy environment for the patient, and if predefined thresholds are exceeded, the Arduino platform can take action to adjust the environmental conditions.

Figure 2. Sensor DHT22, Speaker XHD SFB-55 and the relay module.

- Acoustic Alarm for Rapid Alert: A mini speaker XHD SFB-55 is connected to Arduino and is used to emit acoustic warning signals in case the patients clenches their fist. This alarm signal is intended to draw the attention of medical staff or nearby individuals to provide rapid assistance.
- Automatic Fan Control for Patient Comfort: A fan is automatically activated when the patient's arm is flexed, to provide optimal air circulation and maintain a comfortable room temperature. Based on data from the ambient sensor, the Arduino platform can stop or adjust the fan operation to maintain ideal conditions.
- Relay Module with a 5V channel required for air conditioner automation.

By integrating and combining these components with specific functionalities, the application circuit, presented further in Figure 3, was created.

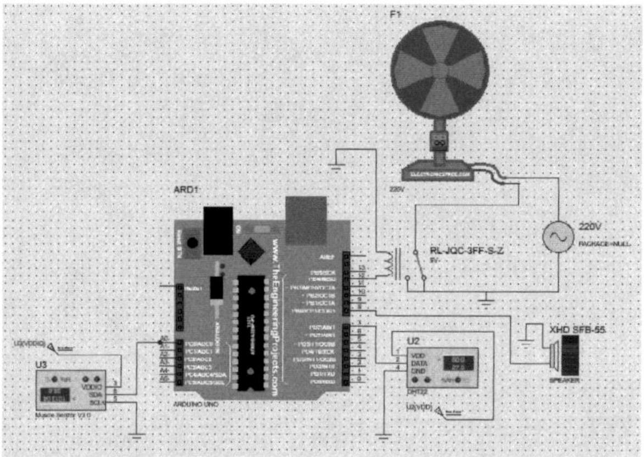

Figure 3. Circuit Diagram of the aplication.

In addition to the hardware components presented earlier, the system also features software functionalities implemented in MATLAB for data processing and advanced control, including:

- Writing MATLAB code for controlling the Arduino platform and processing data acquired by the electromyographic sensors. This code involves initially reading data from the sensors and subsequently, after processing, activating the corresponding devices based on preset conditions.
- The user graphical interface (GUI) is created in MATLAB, allowing medical staff to monitor and interact with the system in real time. The GUI can display sensor data, device status, and other relevant information for patient assistance.

IOP Conf. Series: Materials Science and Engineering **1320** (2024) 012014 doi:10.1088/1757-899X/1320/1/012014

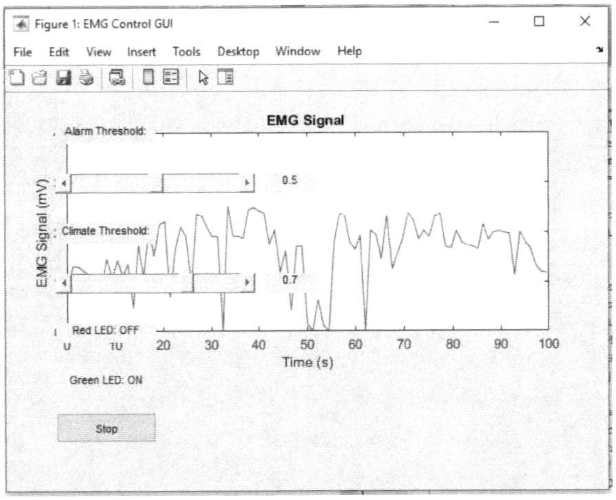

Figure 4. Interface for monitoring the acquired signals.

It can be observed that the corresponding green LED for the air conditioner has turned on, indicating that the medical staff in the guard room has been alerted regarding the parameters in the ward.

Therefore, integrating these hardware and software components provides a versatile solution for monitoring and assisting patients with ischemic stroke. The use of the Arduino platform and the MATLAB environment enables the easy implementation of complex functionalities, contributing to improving the quality of life for patients and the efficiency of medical care.

4.2 Describing the functionality of the application based on electromyographic signal acquisition

To ensure proper data acquired and to avoid interference between the alarm command and the air conditioner command, the application is structured to operate over a 100-second interval, divided into 10 sequences of 10 seconds each. This division into sequences allows for efficient management of actions and accurate data collection at each stage of the process.

During the first 10-second sequence, the electromyographic (EMG) signal is monitored to detect muscle activity associated with fist clenching. This sequence is exclusively dedicated to triggering the alarm in the case when significant muscle activity is detected. In the second 10-second sequence, the EMG signal is evaluated to detect arm flexion, indicating the need to activate the air conditioner for the patient's thermal comfort. Thus, it is planned that in the odd sequences, the alarm will be activated, while in the even-numbered ones, the air conditioner will be started according to the recorded signal. These sequences run, succeed each other, so that in odd-order sequences EMG signals for fist clenching, (if present), will be acquired, while in even-order sequences signals, for arm flexion are acquired.

It is important to mention that the alarm's operating duration is limited to 30 seconds to avoid potential interference or discomfort for the patient. Additionally, the air conditioner will automatically stop when the set temperature and humidity thresholds are reached, ensuring an optimal environment for the patient. By structuring the application into these discrete sequences and implementing time limits for specific actions, efficient system operation is ensured, along with appropriate use in the care and recovery of ischemic stroke patients.

IOP Conf. Series: Materials Science and Engineering **1320** (2024) 012014 doi:10.1088/1757-899X/1320/1/012014

4.2.1 Signal acquisition test

In the first stage, surface electrodes are applied to the relevant arm muscles, such as flexor and extensor muscles. Then, the application initiates the acquisition of the EMG signal from these electrodes using electromyographic sensors. The patient is instructed to perform specific movements, initially to clench the fist and then to relax the arm.

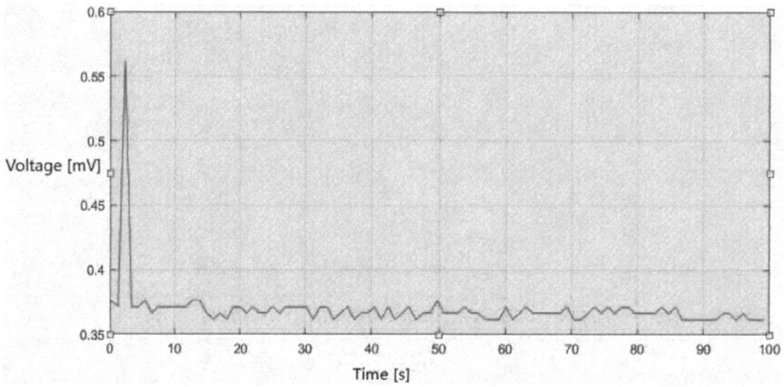

Figure 5. EMG recording signal

We can observe in the previous figure an integrated EMG signal. The recording shows a peak in the first sequence of 0.55 mV, amplitude corresponding to the fist clenching, followed by a simple signal path with an amplitude ranging between 0.36 mV and 0.38 mV during arm relaxation.

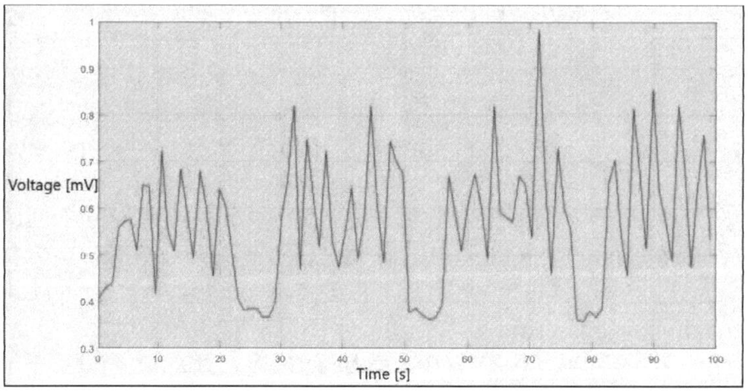

Figure 6. EMG signal recorded during arm flexion and return movement .

Examining Figure 6, we see an integrated EMG signal over a duration of 100 seconds obtained through arm flexion. For the arm flexion and return movement, a potential difference of 0.2 mV was recorded, as shown in Figure 6. Four cycles of arm flexion are monitored, with an approximate medium peak potential of 0.75 mV.

4.2.2 Calculating the average values of the electric potential:
The obtained EMG signals obtained are processed by the application, by calculating the average values of the electric potential during the arm relaxation phases. These average values are used as a reference for comparison with EMG signals recorded during other movements, such as fist clenching or arm flexion. Therefore, the reference potential values for the alarm is considered to be 0.5 mV and for activating the fan 0.7 mV, for each patient.

4.2.3 Triggering the alarm upon fist clenching and starting the ventilation when the arm is flexed:
The application continuously monitors the EMG signals and compares their level with the established average values during the arm relaxation phase. When the signals indicate a significant increase in muscle activity (clenching of the fist), the application triggers the alarm siren to alert the medical staff. When the EMG potentials, as measured by their amplitude, indicate arm flexion, the application sends a command to the air conditioner to start and provide thermal comfort to the patient.

5. Conclusions

The developed application demonstrates significant effectiveness for patients with contralateral hemiplegia following an ischemic stroke, allowing the monitoring of the unaffected arm, thus having a significant impact on the autonomy and quality of life of these patients.

By providing constant feedback and real-time assistance, the application can improve patients' ability to manage their daily activities and gradually regain partial independence. Feedback obtained from patients and medical staff is essential in evaluating the ease of use and utility of the application.

At this time, the application cannot be used to estimate the degree of disability of a patient suffering from an ischemic stroke, nor can it estimate their recovery. However, in the future, further development and improvement of the application could involve integrating additional functionalities, such as electromyographic monitoring of paralyzed limbs to highlight progress following specific rehabilitation procedures.

References

[1] Feigin VL, Norrving B, Mensah GA. Global Burden of Stroke. Circ Res. 2017;120(3):439-448. doi:10.1161/CIRCRESAHA.116.308413
[2] Benjamin EJ, Muntner P, Alonso A, et al. Heart Disease and Stroke Statistics-2019 Update: A Report From the American Heart Association. Circulation. 2019;139(10):e56-e528. doi:10.1161/CIR.0000000000000659
[3] Dobkin BH. Rehabilitation After Stroke. N Engl J Med. 2005;352(16):1677-1684. doi:10.1056/NEJMcp043511
[4] Laver KE, Lange B, George S, Deutsch JE, Saposnik G, Crotty M. Virtual reality for stroke rehabilitation. Cochrane Database Syst Rev. 2017;11(11):CD008349. doi:10.1002/14651858.CD008349.pub4
[5] Merletti, R., & Parker, P. A. (2004). Electromyography: Physiology, engineering, and noninvasive applications. John Wiley & Sons
[6] Delagi, E. F., Perotto, A., Iazzetti, J., Morrison, D., & Stephens, M. (2005). Anatomical guide for the electromyographer: The limbs and trunk. Charles C Thomas Publisher
[7] Adi, A. N (2010). Mechatronics, Yogyakarta: Science Graha
[8] Basmajian, J. V., & De Luca, C. J. (1985). Muscles alive: Their functions revealed by electromyography. Williams& Wilkins
[9] Criswell, E. (2010). Cram's Introduction to Surface Electromyography. Jones & Bartlett Publishers
[10] Langhorne P, Coupar F, Pollock A. Motor recovery after stroke: a systematic review. Lancet Neurol, 2009; 8:741-754. 2:10.16/S1474-4422(09)70150-4
[11] Dobkin BH.Rehabilitation After Stroke. N Engl J Med. 2005;352(16):1677-1684.doi:10.1056/NEJMcp043511

[12] GBD 2019 Diseases and Injuries Collaborators. Global burden of 369 diseases and injuries in 2040 and projections to 2040: an updated systematic analysis for the Global Burden of Disease Study 2019. Lancet. 2020;396(10258):1204-1222. doi:10.1016/S0140-6736(20)30925-9

[13] Donnan GA, Fisher M, Macleod M, Davis SM. Stroke. Lancet. 2008;371(9624):1612-1623. doi:10.1016/S0140-6736(08)60694-7

[14] Benjamin EJ, Blaha MJ, Chiuve SE, et al. Heart Disease and Stroke Statistics-2017 Update: A Report From the American Heart Association. Circulation. 2017;135(10):e146-e603. doi:10.1161/CIR.0000000000000485

[15] Krishnamurthi RV, Ikeda T, Feigin VL. Global, Regional and Country-Specific Burden of Ischaemic Stroke, Intracerebral Haemorrhage and Subarachnoid Haemorrhage: A Systematic Analysis of the Global Burden of Disease Study 2017. Neuroepidemiology. 2020;54(2):171-179. doi:10.1159/000504160

[16] Mihai L, Dudau A, Amariei DE, Lupulescu IS. Arduino-controlled robotic arm for stroke rehabilitation. In: 2017 21st International Conference on System Theory, Control and Computing (ICSTCC). IEEE; 2017:706-710. doi:10.1109/ICSTCC.2017.8106938

[17] Dorsch AK, Thomas S, Xu X, Kaiser W, Dobkin BH. Design and Evaluation of a Wearable Soft Robotic Device for Ankle Rehabilitation After Stroke. Front Neurosci. 2019;13:918. doi:10.3389/fnins.2019.00918

[18] Zahiu Denise; (2019) Physiology of the muscular system, Didactic Publishing House

IOP Conf. Series: Materials Science and Engineering **1320** (2024) 012015 doi:10.1088/1757-899X/1320/1/012015

Convergence of science and technology for wellbeing: A case study on AAL systems for older adults

Joaquín Roca-González[1*], Francisco J. Ortiz-Zaragoza[1], José A. Vera-Repullo[1] and Inmaculada Méndez-Mateo[2]

[1] Dpt. Automation, Electrical Engineering and Electronics Technology, Universidad Politécnica de Cartagena (UPCT), Cartagena, Spain

[2] Dpt. Developmental and Educational Psychology, Universidad de Murcia, Murcia, Spain

[*]E-mail: jroca.gonzalez@upct.es

Abstract. This paper explores the convergence of science and technology through the lens of Ambient Assisted Living (AAL) systems, demonstrating how collaborative development across multiple disciplines enhances the wellbeing and independence of older adults. We illustrate the integration of medical sciences, robotics, artificial intelligence (AI), sensor technology, and human-computer interaction (HCI) in creating sophisticated AAL systems that address both the medical and social needs of older adults. This case study focuses on the principles of convergence that guide the development of modern healthcare solutions, such as as holistic views, common goals, evolutionary patterns, and cross-domain interactions that foster the creation of innovative, effective, and ethically sound solutions. By examining collaborative processes, scientific innovations, and technological advancements, the study highlights the multifaceted impacts of AAL systems on elderly wellbeing, emphasizing the critical role of interdisciplinary approaches in solving complex health and social challenges.

1. Introduction

The transformative convergence of science and technology has consistently redefined the boundaries of human capability and knowledge. Convergence facilitates the amalgamation of diverse scientific and technological trajectories, potentially yielding outcomes that surpass the aggregate effects of their individual contributions requiring of the establishment of dedicated environments explicitly designed for collaborative, transdisciplinary research to achieve higher-order common objectives across the involved disciplines [1]. Today, this convergence is particularly impactful in the domain of healthcare, where interdisciplinary innovations are shaping the future of medical treatments and patient care [2]. This shift towards a more integrated and technologically advanced healthcare landscape is crucial in addressing some of the most pressing challenges faced by society, particularly in the care of older adults. This work will navigate through historical and contemporary perspectives, illustrating how the fusion of diverse scientific knowledge and technological expertise has culminated in the creation of sophisticated systems that not only cater to the medical needs of older adults but also address their psychological and social wellbeing.

This integrative approach is vital in constructing supportive environments that promote an enhanced sense of independence and a higher quality of life among older adults. This work will outline the principles of convergence that guide modern healthcare development, as identified by Roco [3]. It explores how strategic collaboration and technological integration lead to transformative healthcare outcomes. Ambient Assisted Living (AAL) systems [4] serve as a prime example of integrating multiple scientific and technological disciplines to improve elderly wellbeing. These systems mark a significant shift in care delivery, enhancing life quality by seamlessly integrating advanced technologies into everyday living environments.

The development of AAL systems showcases the efficacy of modern engineering and medical sciences in promoting independent living among aging populations. This study will delve into the collaborative processes, scientific innovations, and technological advancements behind AAL systems, highlighting their comprehensive impact on elderly wellbeing. It underscores the critical role of interdisciplinary approaches in addressing complex health and social challenges.

Moreover, this work seeks to inspire ongoing innovation and collaboration, emphasizing the significant societal benefits of such efforts. By examining the interplay between technology and human needs through this case study, we aim to illustrate a future where technology and compassion merge, enhancing the dignity and quality of life for all.

2. Knowledge, scientific research and medical technology evolution

The relentless pursuit of improved health outcomes has driven the evolution of medical technology over millennia. From rudimentary tools to sophisticated digital systems, the scope of medical innovation reflects a deep intertwining of technology with health sciences, leading to enhanced diagnostics, treatments, and patient care. The evolution of medical technology is a testament to humanity's enduring quest for better health and longevity [5][6][7]. Each era's contributions have built upon past knowledge, propelling medicine into new realms of possibility and improving life quality across generations.

The end of the XXth century marked the rise of biotechnology and the beginning of mass collaborative research thanks to the advent of supercomputing capabilities and the spread of internet. Scientific journals and experimental data previously shared in digital media (floppy disks, CD-ROMs, DVDs, etc.) are now accessible by the scientific community across the globe in seconds. This period is characterized by an intense collaboration among thematic research groups at international level fostered by both technological resources and public research policies.

3. Convergence of science and technology in biomedical engineering

Modern medical technology is characterized by digital innovations and artificial intelligence, enhancing surgical precision and enabling personalized medicine, which promises transformative changes in healthcare delivery and effectiveness. The 21st century is recognized as the Era of Convergence [8], where the seamless integration of various technologies and disciplines is used to enhance human potential and address complex global challenges.

Research areas, such as cancer treatment and diagnosis, require tackling high levels of complexity with limited information. Traditional single-domain approaches are inadequate for these challenges, necessitating a multidisciplinary problem-solving strategy known as convergence.

This approach is guided by seven key principles [3] that aim to transform systems by leveraging holistic views, common goals, evolutionary patterns, system-centric actions, cross-domain languages, multitasking, and added-value resource confluence:

I. Holistic View: Understanding and leveraging the interconnectedness and unity in nature and society. Methods include integrating different information systems and altering interactions within and between domains to influence system outcomes.

II. Common Goal: Setting and achieving visionary goals that address common system challenges. It involves using research and innovation to forecast future scenarios and prepare people, tools, and infrastructure for emerging technologies and relationships.

III. Evolution Pattern: Natural and human processes typically follow a cycle of convergence and divergence driven by innovation over time. This principle involves understanding the spiral path of development across different domains, influenced by internal mechanisms and external factors.

IV. System-Centric Actions: Decision-making should consider the entire system and its evolution, combining top-down system vision with bottom-up research input, along with lateral and temporal effects, to make informed decisions in complex hierarchical systems.

V. Cross-Domain Languages: Creating and applying languages that span multiple fields facilitates knowledge transfer, synergy, and innovative solutions. This includes using universal languages such as mathematics and computer code, and sharing large multidomain databases to foster multidisciplinary research and education.

VI. Multi-Tasking: Addressing concurrent cause-and-effect pathways in large systems requires multitasking using various observation angles, methods, and tasks to reach system goals and tools for knowledge mapping and network visualization.

VII. Added-Value: Effective convergence of resources leads to significant and rapid system changes. This principle focuses on the proper coordination of resources and transformative actions, driven by scientific progress and external collaborations to achieve compelling goals.

Thus, convergence facilitates the amalgamation of diverse scientific and technological trajectories, potentially yielding outcomes that surpass the aggregate effects of their individual contributions requiring of the establishment of dedicated environments explicitly designed for collaborative, transdisciplinary research to achieve higher-order common objectives across the involved disciplines. Convergence of science and technology in the medical device field is epitomized by recent advancements that underscore the critical importance of interdisciplinary collaboration in addressing complex healthcare challenges. This section highlights seven pivotal examples of how integrated scientific and technological efforts are reshaping cancer research, diagnostics, personalized medicine, and more.

- Cancer Research: The imperative for converging science and technology in cancer diagnosis and treatment is underscored by the intricate nature of the disease and the voluminous data and knowledge that now exceed the capabilities of traditional interdisciplinary research [9]. Cancer's complexity, characterized by heterogeneous genetic mutations, varying molecular pathways, and diverse patient responses to treatments, necessitates a multifaceted approach

IOP Conf. Series: Materials Science and Engineering **1320** (2024) 012015 doi:10.1088/1757-899X/1320/1/012015

that integrates advancements from multiple scientific and technological fields. Recent breakthroughs in genomics, proteomics, and bioinformatics have generated extensive datasets that require sophisticated computational tools for effective analysis and interpretation. The application of artificial intelligence (AI) and machine learning (ML) in processing and analyzing these large-scale datasets has proven essential for identifying patterns and predicting disease trajectories that are not apparent through conventional methods. The convergence of disciplines not only accelerates the pace of discovery but also enhances the translation of these innovations into clinical applications. By fostering collaboration between fields such as computational biology, clinical oncology, genetics, chemistry, and engineering, we can create a synergistic research environment capable of tackling the multifaceted challenges of cancer.

- Lab-on-chip (LOC) medical devices: The convergence of science and technology in the development of lab-on-chip (LOC) applications is transforming biomedical research, diagnostics, and personalized medicine [10]. LOC technology integrates microfluidics, nanotechnology, and biosensing into miniaturized platforms, enabling complex laboratory procedures on a micro-scale chip. This interdisciplinary approach offers several advantages, including reduced sample volumes, faster processing times, and enhanced assay sensitivity and specificity. The development and application of LOC technology necessitate a multidisciplinary approach, involving microengineering, chemistry, biology, and computer science expertise. Besides, collaborative efforts are essential to address challenges related to fabrication, scalability, and integration of LOC devices into existing healthcare systems.

- Medical Imaging: Developments across computer science, engineering, physics, and biology have led to the creation of sophisticated imaging modalities that offer unparalleled precision, resolution, and diagnostic capabilities. This interdisciplinary synergy has enhanced the effectiveness of techniques such as magnetic resonance imaging (MRI), computed tomography (CT), positron emission tomography (PET), and ultrasound. Significant technological advancements, notably in artificial intelligence (AI) and machine learning (ML), have improved the acquisition, reconstruction, and interpretation of imaging data [11]. AI algorithms are adept at rapidly analyzing large volumes of imaging data, thereby enhancing diagnostic accuracy and facilitating the early detection of diseases. ML models, trained on extensive datasets, are instrumental in detecting subtle patterns and anomalies that may escape human observation, thus supporting more precise and tailored patient care. The development and implementation of these advanced medical imaging devices necessitate a collaborative, interdisciplinary approach involving experts from various scientific (medicine, biology, physics, chemistry, mathematics and statistics, data science) and engineering (biomedical, computer science, electrical, electronics, mechanical, materials, nanotechnology) domains. Such collaboration is crucial for addressing challenges in device miniaturization, data integration, and regulatory compliance, ensuring that these technological innovations are effectively translated into clinical practice.

- 3D Printing in Medicine: 3D printing and bioprinting integrate principles from materials science, bioengineering, and computer-aided design to fabricate complex biological structures, which can be used in various medical applications, including tissue engineering, regenerative medicine, and the creation of anatomical models for surgical planning [12]. 3D printing technology has been instrumental in developing customized prosthetics and implants tailored to the specific anatomical and physiological requirements of patients, enabling the compatibility and functionality of medical implants, leading to improved patient outcomes. Bioprinting extends the capabilities of traditional 3D printing to include the processing of biomaterials and living cells, enabling the construction of tissue constructs with functional biological properties by layering living cells along with biocompatible materials, allowing for the creation of viable tissue structures that closely mimic natural biological systems that hold significant potential for organ fabrication, thus addressing the critical shortage of transplantable organs. The integration of computer modelling and simulation with 3D printing and bioprinting technologies further enhances the design and optimization of medical devices and biological constructs, ensuring that the structures produced are not only biologically functional but also mechanically robust, suitable for their intended medical applications. To fully realize the potential of 3D printing and bioprinting in medicine, a multidisciplinary approach is essential, involving collaboration among specialists in biology, materials science, engineering, and clinical practice.

- Virtual (VR), augmented (AR), and extended reality (XR) in Medicine: The convergence of science and technology through the utilization of virtual reality (VR), augmented reality (AR), and extended reality (XR) is profoundly transforming medical applications, enhancing clinical practices, medical education, and patient care [13]. These immersive technologies integrate advancements from computer science, cognitive psychology, and interactive software engineering to deliver sophisticated tools that are reshaping the landscape of healthcare. The development of these technologies necessitates interdisciplinary collaboration among experts in IT, neurosciences, and clinical disciplines.

- Digital Twins for Healthcare Applications: The development of Digital Twins in healthcare signifies a paradigm shift in medical diagnostics, treatment, and patient care. Digital Twins, which are precise virtual replicas of physical entities, employ advanced technologies such as artificial intelligence, machine learning, and the Internet of Things (IoT) to create comprehensive, dynamic models of patients [14][15]. These models facilitate real-time monitoring and predictive analytics, thereby promoting personalized and proactive healthcare. The implementation of Digital Twins in healthcare necessitates an interdisciplinary approach, combining expertise from fields such as biomedical engineering, computer science, and clinical medicine..

- Robotics in Healthcare: The convergence of science and technology in healthcare robotics [16], is driving significant advancements in diagnostics, treatments, and patient care. These systems enhance precision and reliability in medical procedures, accelerate disease detection through high-throughput screening, and improve laboratory efficiency and accuracy. Effective robotic integration in healthcare requires a multidisciplinary approach that combines expertise from robotics engineering, computer science, medicine, and ethics.

4. Multidisciplinary Convergence in AAL Systems for older adults

The development of systems supporting older adults has emerged as a quintessential example of the convergence of multiple scientific, social and technological disciplines for the creation of holistic and functional environments capable of addressing the complex needs of elderly users.

The aging global population presents significant societal and healthcare challenges, particularly in providing adequate support to elderly individuals who choose to live independently. This demographic shift necessitates innovative solutions to ensure the well-being and health of older adults, particularly those living alone (*II. Common Goal*). AAL systems, offer a promising response to these challenges by merging various scientific and technological disciplines (*I. Holistic View*) by addressing not only the physical and medical needs of the aging population but also provide critical emotional and social support, enhancing their quality of life and extending their ability to live independently. Convergence of science and technology in the development of AAL systems with social robots illustrates a forward-thinking approach to addressing the complex needs of the aging population (*IV. System-Centric Actions*).

4.1Disciplines involved in the design and development of AAL Systems for older adults

By integrating expertise from multiple disciplines such as medicine, psychology, robotics, computer science, healthcare, sensor technology, artificial intelligence (AI) and human-computer interaction and considering the ethical, psychological, and physical aspects of elderly care, these systems promise to enhance the independence and quality of life for older adults living alone. This holistic approach not only enhances the quality of life for elderly individuals but also offers a scalable solution *(III. Evolution Pattern, VII. Added-Value)* that can be adapted to different individuals and living environments, marking a significant step forward in the use of technology to care for our aging society.

- Artificial Intelligence, Computer Science and Robotics: Computer science, especially artificial intelligence (AI) and machine learning (ML), is pivotal in operating social robots, which use these technologies to process speech, recognize faces and emotions, and adapt their behavior through interaction [17]. Robotics integrates necessary hardware and software for navigating spaces and facilitating physical and social interactions. Sensor technology and actuation developments enable robots to function safely within home settings [18]. In ambient assisted living environments, these social robots offer companionship [19] and practical support, such as medication reminders and emergency responses, by utilizing advanced AI to sensitively react to users' emotional and physical needs [20].

- Biomedical Engineering and Sensor Technology: Biomedical engineering contributes through the development of sensors and devices that can monitor vital signs such as heart rate, blood pressure, and blood glucose levels non-invasively using wearable unobtrusive and effective devices, providing real-time data that allows for regular monitoring of older adults's health status, alerting healthcare providers to potential issues before they become critical and facilitate remote consultations and health assessments without the need for frequent hospital visits. Embedded sensors in the environment and wearable devices collect data on the user's physical, emotional [21] and environmental conditions [22] which are required for creating a responsive AAL system that adapts to the user's changing needs, ensuring safety, comfort and well-being.

- Data Science and Big Data Analytics: The vast amounts of data generated by sensors and wearables are processed and analyzed using big data analytics. Data scientists work to develop algorithms that can predict health deteriorations or falls, and changes in the emotional state of users, enabling preemptive medical and psychological interventions.

- Human-Computer Interaction (HCI): HCI specialists focus on making AALS intuitive and user-friendly for elderly users, who may not be tech-savvy. This includes the design of interfaces and interaction modalities that are easy to understand and operate [23].

- Healthcare and Medicine: Medical professionals, including gerontologists and primary care providers, are integral to the development of AALS. They provide essential insights into the health needs and challenges faced by older adults. This medical expertise ensures that the technologies developed are aligned with practical healthcare applications, such as disease management and rehabilitation.

- Psychology and Social Work: Understanding the psychological aspects of aging and isolation is crucial for programming robots that can effectively engage and interact with elderly users, so psychology and social work specialist guide the design of robot interactions to ensure they are comforting. These disciplines help define the behavioral strategies robots should adopt to promote mental well-being and reduce feelings of loneliness and social isolation [24].

- Implementation and User Acceptance: User acceptance is critical for the successful deployment of AALS. Studies involving elderly participants help designers understand the preferences and challenges faced by end-users, leading to better system designs. Social scientists and anthropologists play a role in studying these interactions and feedback to refine the technologies further [25].

- Ethics: Ethical considerations are paramount in the deployment of AAL systems, since the integration of these technologies into the lives of elderly individuals also raises significant ethical issues, such as privacy, autonomy, and consent. Ethicists ensure that the systems operate within an ethical framework that respects the dignity and privacy of users by securely handling personal data and ensuring that interactions do not undermine the autonomy or self-esteem of older adults [26]. Ethicists and legal experts need to collaborate closely with technologists and healthcare providers to develop guidelines that protect the rights and dignity of users.

4.2 Stakeholders involved in collaborative development of AAL systems for older adults

The development and operation of AAL systems involve a wide array of stakeholders, each contributing different perspectives and expertise:

- Researchers and Developer: Typically, from fields like AI, robotics, biomedical engineering, and gerontology, these professionals drive the technological and conceptual advancements in AAL systems develop and refine the technical aspects of AALS, including robotics, wearable devices, and sensor networks.

- Designers and HCI Experts: Work to ensure that the technology is accessible and acceptable to elderly users.

- Data Scientists: Analyze and interpret the data to provide actionable insights for both users and caregivers.

- Healthcare Professionals: Doctors, nurses, therapists and caregivers provide insights into the medical needs of older adults and emergency response strategies, helping tailoring the AAL systems to improve health outcomes and ensuring that the systems offer meaningful social interactions and emotional support [27].

- End Users and Their Families: Their feedback is crucial for refining the system's design and functionality, ensuring that it meets their actual needs and preferences.

- Policy Makers and Regulators: They ensure that the systems meet safety and privacy regulations and help integrate these technologies into broader health and social policies.

- Ethicists: Guiding the development and deployment processes to align with ethical standards, societal values, norms and expectations addressing privacy, consent, and autonomy related issues.

4.3 Co-Creation workflow for collaborative development of AAL systems for older adults

Developing an ambient living system with wearables and social robots to support the health and emotional state of elderly people involves a comprehensive, multi-staged process *(VI. Multi-Tasking)* engaging a diverse range of stakeholders. This workflow outlines the procedures from the initial idea to the commercialization of the final product under a co-creation scheme *(V. Cross-Domain Languages)* involving engineers, scientists, physicians, psychologists, caregivers, users, and their families:

a) Ideation and Conceptualization: Engage with elderly users, caregivers, and healthcare providers to identify specific needs and challenges and perform interviews, surveys, and observation sessions as required for needs assessment. Converge ideas from stakeholders *(I. Holistic View)*, including potential technology applications and solutions (e.g., types of sensors for wearables, functionalities for social robots) for concept development. Engineers and scientists analyze technical feasibility, while market analysts assess economic viability *(VI. Multi-Tasking, VII. Added-Value)*.

b) Design and Prototyping: Organize collaborative sessions as interdisciplinary workshops where engineers, designers, and healthcare professionals brainstorm and sketch out system designs *(V. Cross-Domain Languages)*. Engineers and technologists develop initial prototypes of wearables and robots, incorporating input from psychologists to ensure they address emotional and cognitive aspects. Involve elderly users and caregivers in testing prototypes as user feedback loops *(III. Evolution Pattern)* to provide information required to refine the designs.

c) Technical Development and Integration: Technologists and engineers integrate various components (wearables, robots, and ambient sensors) to function seamlessly as a whole system. Computer scientists and software engineers develop and test software algorithms for data collection, analysis, and user interaction interfaces. Safety and compliance are performed to ensure all components meet health, safety, and regulatory standards *(VI. Multi-Tasking)*.

d) Pilot Testing and Validation: Implement the system in a controlled group of households to monitor functionality in real-world tests and gather extensive data *(III. Evolution Pattern)*. Conduct clinical trials to validate the health benefits and emotional support capabilities of the system, involving physicians and psychologists. Refine the system based on feedback and trial outcomes, focusing on user comfort, system reliability, and effectiveness *(IV. System-Centric Actions)*.

e) Production and Quality Assurance: Transition from prototype to mass production, as required to scale up production, setting up supply chains and manufacturing processes. Implement rigorous quality assurance testing phases to ensure the products meet all specifications and standards *(VII. Added-Value)*.

f) Marketing and Commercialization: Develop marketing strategies and plans targeting specific demographics, highlighting the benefits and unique selling propositions of the ambient living system. Establish distribution partnerships and channels, including online platforms, healthcare providers, and senior living communities. Set up customer service and technical support to assist users and caregivers with system operation and troubleshooting *(III. Evolution Pattern, VII. Added-Value)*.

g) Post-Launch Monitoring and Continuous Improvement: Continuously collect user feedback on the system's performance and impact on their quality of life. Keep refining the product based on new and ongoing scientific findings and technological advancements. Maintain an ongoing dialogue with all stakeholders to ensure the system evolves in line with user needs and technological possibilities *(III. Evolution Pattern, IV. System-Centric Actions, VII. Added-Value)*.

Throughout these stages, the co-creation approach ensures that the voices of all stakeholders, especially the end users and their families, are considered *(V. Cross-Domain Languages)*. This methodology not only enhances the functionality and usability of the developed system but also fosters acceptance and satisfaction among elderly users *(VII. Added-Value)*, ultimately leading to better health and emotional well-being outcomes *(II. Common Goal)*.

5. Conclusion

The interaction between various scientific fields and technological innovations has ushered in new medical possibilities. Tracing the evolution of medical technology from ancient practices to modern digital innovations, it is clear that the advancement of patient care has consistently depended on merging diverse knowledge and techniques.

This historical perspective not only underscores the persistent human effort to address health challenges but also enhances our understanding of the complexities involved in modern medical technology, especially in elderly care.

The convergence of science and technology in the development of Ambient Assisted Living (AAL) systems provides a critical case study in enhancing the well-being of older adults. This exploration reveals how such convergence revolutionizes the development and effectiveness of medical technologies tailored to the complex needs of the aging population. By employing a multi-disciplinary and collaborative approach, AAL systems demonstrate the profound benefits of integrating diverse scientific knowledge and technological advancements to improve the quality of life and independence of elderly individuals.

AAL systems exemplify the strategic synergy of robotics, AI, sensor technology, and healthcare expertise, addressing the multifaceted challenges of aging. These systems are crafted through collaborative efforts, emphasizing the need for technological adaptability and user-centered design to meet the physiological, psychological, and social needs of older adults. The guiding principles of convergence in AAL development—holistic views, system-centric actions, cross-domain languages, and multitasking—establish a strong framework for innovation. These principles ensure that AAL systems are not only technologically advanced but also culturally and ethically sound, meeting the highest standards of medical ethics and patient care. The case study discussed highlights the essential role of multidisciplinary collaboration involving academics, researchers, healthcare professionals, and industry experts. This collaboration is crucial for creating a conducive environment for the practical application of AAL technologies, with the involvement of multiple stakeholders, including engineers, ethicists, and end-users, to refine these systems for effectiveness and acceptability.

As we progress further into the 21st century, the fusion of science and technology will increasingly address the challenges of global aging populations. The ongoing advancement of AAL systems represents a significant step towards leveraging technological innovation for social benefit, particularly in enhancing elderly well-being. The insights from this case study not only contribute to academic and practical knowledge but also underscore the necessity for continuous investment and research in this crucial area of public health and technology.

In conclusion, the collaborative creation of AAL systems through the lens of scientific and technological convergence offers a vision of the future where technology profoundly serves humanity's needs, making life not only longer but also more fulfilling and independent. This endeavor challenges us to reconsider the boundaries between disciplines and to reimagine the potential of integrated solutions that are as compassionate as they are innovative.

References

[1] Roco M C and Bainbridge W S 2003 *Converging Technologies for Improving Human Performance* (Dordrecht: Springer Netherlands)

[2] Sharp P, Jacks T and Hockfield S 2016 Capitalizing on convergence for health care *Science* **352** 1522–3

[3] Roco M C 2020 Principles of convergence in nature and society and their application: from nanoscale, digits, and logic steps to global progress *J Nanopart Res* **22** 321

[4] Memon M, Wagner S R, Pedersen C F, Beevi F H A and Hansen F O 2014 Ambient Assisted Living Healthcare Frameworks, Platforms, Standards, and Quality Attributes *Sensors* **14** 4312–41

[5] Cruse J M 1999 History of Medicine: The Metamorphosis of Scientific Medicine in the Ever-Present Past *The American Journal of the Medical Sciences* **318** 171–80

[6] Neuman M R, Baura G D, Meldrum S, Soykan O, Valentinuzzi M E, Leder R S, Micera S and Zhang Y-T 2012 Advances in Medical Devices and Medical Electronics *Proceedings of the IEEE* **100** 1537–50

[7] Cutler D M and McClellan M 2001 Is Technological Change In Medicine Worth It? *Health Affairs* **20** 11–29

[8] Bainbridge W and Roco M C 2016 Handbook of science and technology convergence (New York, NY: Springer Berlin Heidelberg)

[9] Iacobuzio-Donahue C A 2024 Supporting Convergence Science Cancer Research 84 947–9

[10] Cooper J M 2022 Challenges in lab-on-a-chip technology *Front. Lab. Chip. Technol.* **1**

[11] Panayides A S, Amini A, Filipovic N D, Sharma A, Tsaftaris S A, Young A, Foran D, Do N, Golemati S, Kurc T, Huang K, Nikita K S, Veasey B P, Zervakis M, Saltz J H and Pattichis C S 2020 AI in Medical Imaging Informatics: Current Challenges and Future Directions *IEEE Journal of Biomedical and Health Informatics* **24** 1837–57

[12] Cornejo J, Cornejo-Aguilar J A, Vargas M, Helguero C G, Milanezi de Andrade R, Torres-Montoya S, Asensio-Salazar J, Rivero Calle A, Martínez Santos J, Damon A, Quiñones-Hinojosa A, Quintero-Consuegra M D, Umaña J P, Gallo-Bernal S, Briceño M, Tripodi P, Sebastian R, Perales-Villarroel P, De la Cruz-Ku G, Mckenzie T, Arruarana V S, Ji J, Zuluaga L, Haehn D A, Paoli A, Villa J C, Martinez R, Gonzalez C, Grossmann R J, Escalona G, Cinelli I and Russomano T 2022 Anatomical Engineering and 3D Printing for Surgery and Medical Devices: International Review and Future Exponential Innovations *BioMed Research International* **2022** e6797745

[13] Torner J, Skouras S, Molinuevo J L, Gispert J D and Alpiste F 2019 Multipurpose Virtual Reality Environment for Biomedical and Health Applications *IEEE Transactions on Neural Systems and Rehabilitation Engineering* **27** 1511–20

[14] Barricelli B R, Casiraghi E and Fogli D 2019 A Survey on Digital Twin: Definitions, Characteristics, Applications, and Design Implications *IEEE Access* **7** 167653–71

[15] Katsoulakis E, Wang Q, Wu H, Shahriyari L, Fletcher R, Liu J, Achenie L, Liu H, Jackson P, Xiao Y, Syeda-Mahmood T, Tuli R and Deng J 2024 Digital twins for health: a scoping review *npj Digit. Med.* **7** 1–11

[16] Jovanovic K, Schwier A, Matheson E, Xiloyannis M, Rozeboom E, Hochhausen N, Vermeulen B, Graf B, Wolf P, Nawrat Z, Escuder J, Mechelinck M, Sorensen B, Boscolo P R, Obach M, Tognarelli S, Jankovic M, Leroux C, Ferrigno G, Siepel F J and Stramigioli S 2021 Digital Innovation Hubs in Health-Care Robotics Fighting COVID-19: Novel Support for Patients and Health-Care Workers Across Europe *IEEE Robotics & Automation Magazine* **28** 40–7

[17] Bohr A and Memarzadeh K 2020 Chapter 2 - The rise of artificial intelligence in healthcare applications *Artificial Intelligence in Healthcare* ed A Bohr and K Memarzadeh (Academic Press) pp 25–60

[18] Barber R, Ortiz F J, Garrido S, Calatrava-Nicolás F M, Mora A, Prados A, Vera-Repullo J A, Roca-González J, Méndez I and Mozos Ó M 2022 A Multirobot System in an Assisted Home Environment to Support the Elderly in Their Daily Lives *Sensors* **22** 7983

[19] Calatrava-Nicolás F M, Gutiérrez-Maestro E, Bautista-Salinas D, Ortiz F J, González J R, Vera-Repullo J A, Jiménez-Buendía M, Méndez I, Ruiz-Esteban C and Mozos O M 2021 Robotic-Based Well-Being Monitoring and Coaching System for the Elderly in Their Daily Activities *Sensors* **21** 6865

[20] Toupas P, Tsamis G, Kargakos A, Giakoumis D, Votis K and Tzovaras D 2023 A Framework Towards Ambient Assisted Living Enhanced by Service Robots *Proceedings of the 16th International Conference on PErvasive Technologies Related to Assistive Environments* PETRA '23 (New York, NY, USA: Association for Computing Machinery) pp 423–8

[21] Bautista-Salinas D, González J R, Méndez I and Mozos O M 2019 Monitoring and Prediction of Mood in Elderly People during Daily Life Activities *2019 41st Annual International Conference of the IEEE Engineering in Medicine and Biology Society (EMBC)* 2019 41st Annual International Conference of the IEEE Engineering in Medicine and Biology Society (EMBC) pp 6930–4

[22] Nweke H F, Teh Y W, Al-garadi M A and Alo U R 2018 Deep learning algorithms for human activity recognition using mobile and wearable sensor networks: State of the art and research challenges *Expert Systems with Applications* **105** 233–61

[23] Kleinberger T, Becker M, Ras E, Holzinger A and Müller P 2007 Ambient Intelligence in Assisted Living: Enable Elderly People to Handle Future Interfaces Universal Access in Human-Computer Interaction. Ambient Interaction ed C Stephanidis (Berlin, Heidelberg: Springer) pp 103–12

[24] Lazarou I, Karakostas A, Stavropoulos T G, Tsompanidis T, Meditskos G, Kompatsiaris I and Tsolaki M 2016 A Novel and Intelligent Home Monitoring System for Care Support of Elders with Cognitive Impairment *J Alzheimers Dis* **54** 1561–91

[25] Ziefle M and Rocker C 2010 Acceptance of pervasive healthcare systems: A comparison of different implementation concepts UCD/PHA WORKSHOP (PervasiveHealth 2010)

[26] Ziefle M, Rocker C and Holzinger A 2011 Medical Technology in Smart Homes: Exploring the User's Perspective on Privacy, Intimacy and Trust *2011 IEEE 35th Annual Computer Software and Applications Conference Workshops* 2011 IEEE 35th Annual Computer Software and Applications Conference Workshops pp 410–5

[27] Hartmann K V, Rubeis G and Primc N 2024 Healthy and Happy? An Ethical Investigation of Emotion Recognition and Regulation Technologies (ERR) within Ambient Assisted Living (AAL) *Sci Eng Ethics* **30** 2

[28] Hartmann K V, Rubeis G and Primc N 2024 Healthy and Happy? An Ethical Investigation of Emotion Recognition and Regulation Technologies (ERR) within Ambient Assisted Living (AAL) *Sci Eng Ethics* **30** 2

IOP Conf. Series: Materials Science and Engineering **1320** (2024) 012016 doi:10.1088/1757-899X/1320/1/012016

Analysis in Hospital Operating Rooms: Understanding Perturbations from Electrocautery

Moruz Matei[1,2], Andritoi Doru[1*] and Ciorap Radu[1]

[1] Biomedical Sciences Department, Faculty of Medical Bioengineering, University of Medicine and Pharmacy "Grigore T. Popa" Iasi, Romania
[2] Faculty of Electrical Engineering, Energetics and Applied Informatics," Gheorghe Asachi" Technical University of Iasi, Romania

*E-mail: andritoi.doru@umfiasi.ro

Abstract. Under normal operation, electronic systems and equipment emit electromagnetic fields (EMFs) at predefined levels. These fields can pose safety risks by inducing currents in conductive components and causing polarization effects on insulators, potentially affecting biological tissue at both cellular and molecular levels. In this study, we used the Narda SRM-3006 magnetic field surveillance equipment to analyse the spatial distribution of magnetic fields in operating rooms and address electromagnetic interference (EMI) concerns. We detailed specific methods for measuring and analysing magnetic fields in hospital operating blocks, including a technique for representing spatial distribution with minimal measurements. Additionally, we aimed to conduct a survey to assess hospital staff's awareness of EMI-related safety hazards. The results of the measurements will assist us in developing best practice guidelines for the use and positioning of equipment and personnel in the operating room to prevent potential accidents.

1. Introduction

Whenever electrical equipment is in operation, it emits an electromagnetic field (EMF), a phenomenon also observed during electricity consumption. The influence of the magnetic field on equipment in the operating room can be significant and potentially hazardous (1). Magnetic fields can interfere with the functioning of electronic devices, such as monitors, anaesthesia machines, and other surgical equipment that rely on electronic components (2).

This interference can lead to malfunction or incorrect readings, posing risks to patient safety and surgical outcomes. To mitigate these risks, operating rooms are typically equipped with guidelines and protocols to minimize the presence of magnetic fields (3). This may include avoiding the use of certain equipment or materials that generate strong magnetic fields, maintaining a safe distance between magnetic sources and sensitive devices, and using shielding or protective measures when necessary (4). Furthermore, healthcare professionals working in the operating room should be trained to recognize and address potential risks associated with magnetic interference to ensure the safety and effectiveness of medical procedures (5). Regular maintenance and testing of equipment are also essential to identify and address any issues related to magnetic interference promptly (6).

IOP Conf. Series: Materials Science and Engineering **1320** (2024) 012016 doi:10.1088/1757-899X/1320/1/012016

The use of the electrocautery is crucial in medicine for surgical procedures and treating tissue lesions. The device utilizes an electric current to coagulate blood and to cut or destroy tissues during surgeries. Its proper use requires adequate knowledge in handling the device to minimize the risks of burns and other complications.
Electrocautery has become an essential tool in modern medical practice, significantly contributing to the efficiency and safety of surgical procedures. There are primarily two types of electric scalpels: monopolar and bipolar.

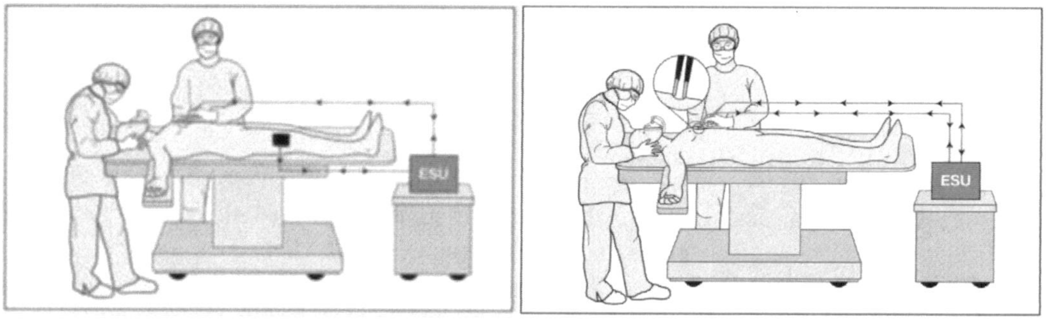

Figure 1. a Monopolar electric scalpel schematic b Bipolar electric scalpel schematic

Schematic, the monopolar electric scalpel (Figure 1a) consists of three components: the electrosurgical generator (ESU), the scalpel itself, and the electrode/dispersal pad. The generator produces high-frequency current conducted through a cable to the metal tip of the electrode, which is fixed on an electrically insulated handle. At the point of contact between this electrode and the tissue, an electric arc is formed. Due to the high energy concentration on a small tissue area, the desired electrosurgical effect is achieved. As the current passes through the patient's tissues to the dispersal pad, its intensity decreases. Due to the large contact area of the dispersal pad with the skin, no thermal effect occurs in the contact area. From the dispersal pad, the current returns to the generator, thus closing the electrical circuit.

The high-frequency generator is operated by the operating surgeon using a pedal or a button on the scalpel handle. The risk of burns at the site of the dispersal pad occurs when it is not properly applied to the patient's skin. Thus, the current intensity over a reduced contact surface can produce an undesired thermal effect, similar to that of the active electrode (electric scalpel).

When the device is designed to apply high-frequency current to two branches of the scalpel through a special clamp (pincer-type) whose branches are insulated from each other, we obtain a so-called "bipolar" effect, characteristic of bipolar scalpels (Figure 1.b). In this case, there is no longer need for a dispersal pad, as the electric arc is formed at the tip of the scalpel, passing only through the tissue between the two tips of the instrument.

The importance of measuring the electromagnetic field in the operating room lies in ensuring the safety of both medical personnel and patients (7)(8). Electromagnetic fields generated by medical equipment, such as electrosurgical devices, can pose health risks if they exceed safe levels (9). Regular measurements help identify any potential hazards and allow for corrective actions to be taken promptly, such as adjusting equipment settings or implementing shielding measures (10).

IOP Conf. Series: Materials Science and Engineering **1320** (2024) 012016 doi:10.1088/1757-899X/1320/1/012016

2. Measurement protocol

The measurement of the electromagnetic field in both configurations is presented in this study. For both monopolar and bipolar configurations, a piece of muscular tissue was used, and the electromagnetic field was measured at multiple points relative to the biological piece, ranging from 50 cm to 210 cm, in total 50 points of measurements (Figure 2). During the measurements, the electrocautery was set at a power of 120 W corresponding to the cutting mode

To assess the electromagnetic field produced by the electrosurgical devices, muscular tissue was positioned on the operating table at a height of 125 cm. Measurements were conducted using an RSA 306B spectrum analyser with a frequency range of 9 kHz to 6.2 GHz, along with an active electric field sensor HZ530 covering 0.1 MHz to 1000 MHz. The analyser was configured in the "+peak" detector mode with "MAX HOLD" enabled to capture and display peak values.

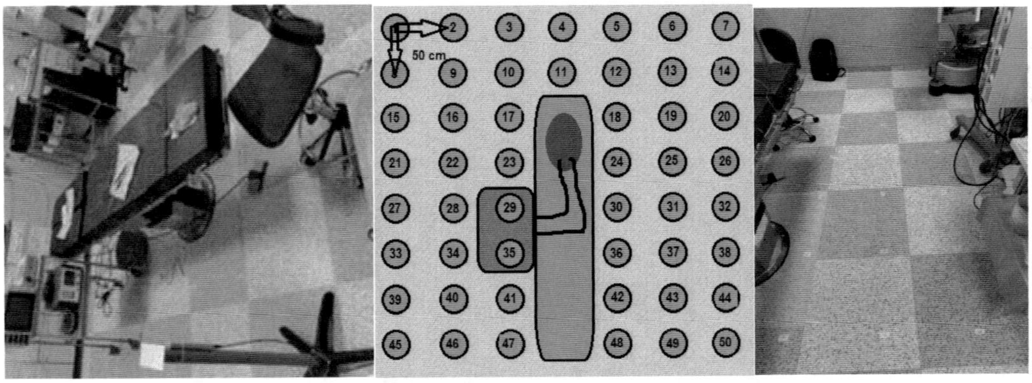

Figure 2. Images from the operating room and the organization of measurement points

Additionally, the resolution bandwidth (RBW) was automatically set to 200 kHz.

Levels of electromagnetic radiation at different frequencies are measured, and the results are recorded. If it is found that the levels exceed established safety limits, corrective measures must be taken, such as adjusting the equipment or redistributing space in the operating room.

3. Results and discussion

The measurements conducted revealed varying levels of electromagnetic fields across different areas of the operating room. Electrosurgical devices produced significant levels of EMF, with notable variations depending on distance and position.

The logarithmic representation of electric field E in dBμV/m provides a visual depiction of the intensity of at various locations. These representations, providing a clearer understanding of both low and high-intensity areas.

Peaks on the graph indicate higher electric field strengths, while valleys suggest lower intensity regions. This graphical representation aids in identifying areas of potential electromagnetic interference and assists in implementing mitigation strategies to ensure the safety and functionality of electronic devices within the environment (Figure 3 and Figure 4).

IOP Conf. Series: Materials Science and Engineering **1320** (2024) 012016 doi:10.1088/1757-899X/1320/1/012016

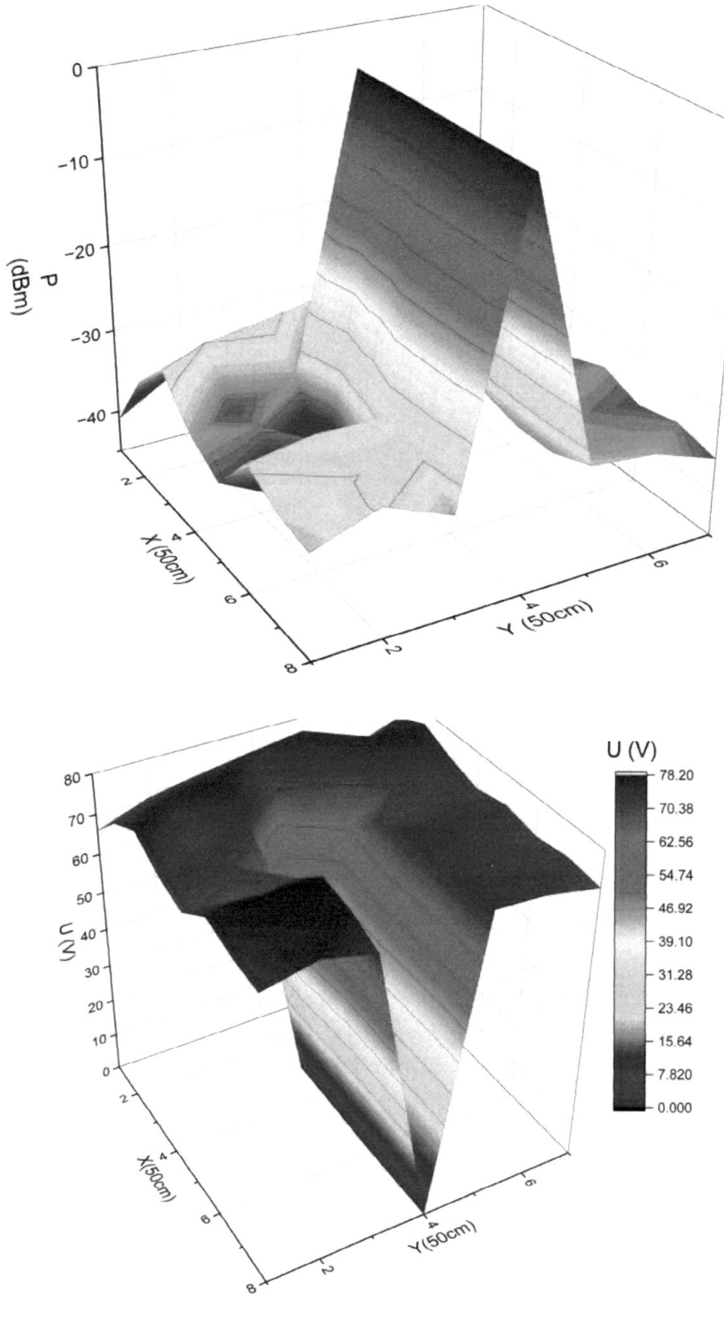

Figure 3. Graphical depiction of power P[dBm] and calculated voltage U[dBμV] based on measurements from each of the 50 spatial points for bipolar mode.

IOP Conf. Series: Materials Science and Engineering **1320** (2024) 012016 doi:10.1088/1757-899X/1320/1/012016

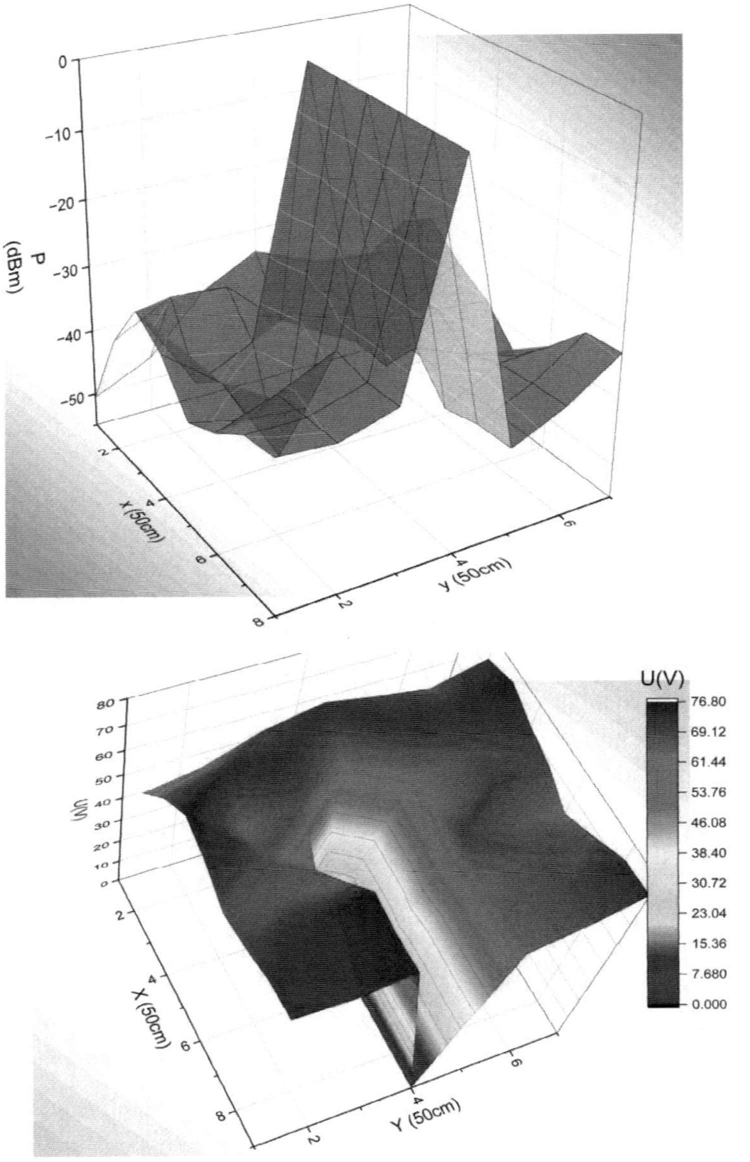

Figure 4. Graphical depiction of power P[dBm] and calculated voltage U[dBμV] based on measurements from each of the 50 spatial points for monopolar mode.

The maximum values recorded were within recommended safety standards, except for some localized amplification points.

It was also observed that an increased intensity value was measured not only in the vicinity of the electrocautery unit but also near the electrodes and cables. This suggests that multiple components of the electrocautery setup contribute to the overall electromagnetic field in the operating room, which is important for understanding potential areas of electromagnetic interference and ensuring the safety of medical personnel and patients.

4. Conclusion

Measuring the electromagnetic field in the operating room is crucial for several reasons. Firstly, it helps ensure the safety of both medical staff and patients by identifying and mitigating potential health risks associated with exposure to electromagnetic radiation.

Several differences have been identified regarding the magnetic field generated in the use of the monopolar mode and the bipolar mode of an electrocautery. In monopolar mode, the electric current flows from the active electrode through the patient's body to the dispersal electrode. This creates a more extensive electromagnetic field that can affect a larger area around the equipment and cables while in bipolar mode, the current flows only between the two tips of the bipolar forceps, confining the electromagnetic field to the immediate area of tissue between these tips. This results in a more localized and concentrated electromagnetic field in the action area. Due to the current passing through a larger area of the patient's body and the use of a dispersal electrode, electromagnetic interference risk is slightly higher in monopolar mode compared to bipolar mode.

All the aforementioned represent the premises of a larger study involving the measurement of these parameters but in multiple configurations that involve a multitude of other equipment that can operate in the operating room (such as a defibrillator) but also the medical staff directly involved in these interventions.

By monitoring the electromagnetic field, healthcare facilities can uphold safety standards, minimize the risk of adverse health effects, and create a secure environment for surgical procedures. Ultimately, the importance of measuring the electromagnetic field lies in creating a safe and efficient environment for surgical procedures, promoting the well-being of everyone involved.

References

[1] Yadin David, W. David Paperman, "Electromagnetic interference in healthcare environment"- Chapter 55, Clinical Engineering Handbook (Second Edition), Academic Press, 2020, Pages 362-372, ISBN 9780128134672, https://doi.org/10.1016/B978-0-12-813467-2.00056-0.

[2] Alnamir H., "Study of Low Frequency Electromagnetic Interference Problems in Hospital Environment," 2019 11th International Symposium on Advanced Topics in Electrical Engineering (ATEE), Bucharest, Romania, 2019, pp. 1-5, doi: 10.1109/ATEE.2019.8724736.

[3] Corciovă C., Andritoi D. and Ciorap R., "Elements of risk assessment in medical equipment," 2013 8TH INTERNATIONAL SYMPOSIUM ON ADVANCED TOPICS IN ELECTRICAL ENGINEERING (ATEE), Bucharest, Romania, 2013, pp. 1-4, doi: 10.1109/ATEE.2013.6563427.

[4] Bhatt, C.R., Redmayne, M., Abramson, M.J. et al. Instruments to assess and measure personal and environmental radiofrequency-electromagnetic field exposures. Australas Phys Eng Sci Med 39, 29–42 (2016). https://doi.org/10.1007/s13246-015-0412-z

[5] Harris, C.; Boivin, W.; Boyd, S.; Coletta, J.; Kerr, L.; Kempa, K.; Aronow, S.. Electromagnetic Field Strength Levels Surrounding Electronic Article Surveillance (EAS) Systems. Health Physics 78(1):p 21-27, January 2000.

[6] Luca C. and Sălceanu A., "Study upon electromagnetic interferences inside an intensive care unit," 2012 International Conference and Exposition on Electrical and Power Engineering, Iasi, Romania, 2012, pp. 535-540, doi: 10.1109/ICEPE.2012.6463878.

[7] Sabra, M., Refaat, M.M. (2024). Handling Pacemakers and Implantable Cardioverter Defibrillators in the Operating Room. In: Hoballah, J.J., Kaafarani, H.M., Tsoulfas, G. (eds) Principles of Perioperative Safety and Efficiency. Springer, Cham. https://doi.org/10.1007/978-3-031-41089-5_8.

[8] Shalaby, M., Shokair, M. & Messiha, N.W. Electromagnetic Field Measurement Instruments: Survey. Iran J Sci Technol Trans Electr Eng 43 (Suppl 1), 1–14 (2019). https://doi.org/10.1007/s40998-018-0116-y.
[9] Peter M. Schulman, Miriam M. Treggiari, N. David Yanez, Charles A. Henrikson, Peter M. Jessel, Thomas A. Dewland, Matthias J. Merkel, Valerie Sera, Izumi Harukuni, Ryan B. Anderson, Ed Kahl, Ann Bingham, Nabil Alkayed, Eric C. Stecker; Electromagnetic Interference with Protocolized Electrosurgery Dispersive Electrode Positioning in Patients with Implantable Cardioverter Defibrillators. Anesthesiology 2019; 130:530–540 doi: https://doi.org/10.1097/ALN.0000000000002571
[10] Stevens, F., Conditt, M.A., Kulkarni, N. et al. Minimizing Electromagnetic Interference from Surgical Instruments on Electromagnetic Surgical Navigation. Clin Orthop Relat Res 468, 2244–2250 (2010). https://doi.org/10.1007/s11999-010-1366-9.

Continuous Transfer Function Modeling for Predicting the COVID-19 Spread

Cristina-Maria Stancioi[1*], Vlad Muresan[1], Mihail Abrudean[1], Mihai Stelian Munteanu[2], Valentin Sita[1]

[1] Automation department, Technical University of Cluj-Napoca, Cluj-Napoca, Romania
[2] Electrical Engineering and Measurements department, Technical University of Cluj-Napoca, Cluj-Napoca, Romania

*E-mail: stancioi_cristina@yahoo.com

Abstract. In this paper, Covid-19, a transmissible virus that is spreading over the globe, represents undoubtedly one of the most profound incidents in human history, influencing people's everyday activities for an extended period of time. Coronaviruses are among the most common infections which harm the human respiratory system. SARS-CoV-2 is mostly transmitted by droplets in the air, especially aerosols, produced by a COVID-19 infected person who breathes in close vicinity to other individuals, sneezes, talks or coughs. The data available at the moment enables significant modelling advancements for the prediction and simulation of the infectious disease dynamic transmission. Quantifying and filtering the collected data is a vital aspect of this research. The current study gives valuable insights for the development of mathematical models for estimating and predicting, with a focus on continuous transfer function models, using data collected from Romanian territory. By taking into account the information received, the structural parameters of the suggested mathematical models are calculated, and the power and effectiveness of the obtained models are tested using simulation.

1. Introduction

The spread of the SARS-CoV-2 (COVID-19) virus has generated an extraordinary global crisis. In this context, science and research are the front line in the fight to stop the spread of the virus and find rapid treatment solutions[1]. The fast evolution, from the first cases to the declaration of the pandemic state on March 11, 2020, focused the attention of the entire scientific and research community around the world[2].

Databases with relevant articles, many of them open source, were created in a short time, with the aim of facilitating specialists' access to specialized articles. And, even if the epidemic caused by COVID-19 interrupted the normal activity of universities and research institutes, the same institutions proposed the same common goal: that through the results of their activity, to stop the spread of the virus and limit as much as possible its effects[3],[4].

Coronaviruses are among the most common infections that damage humans' respiratory systems. SARS-CoV-2 is normally transmitted by respiratory droplets, most particularly aerosols, released by an infected person when sneezing, coughing, talking, singing, or breathing in close contact with others[5]. Exposure can occur by accumulation of respiratory droplets, inhalation of respiratory droplets or aerosol particles, and particles on exposed mucous membranes, and/or or the contact of mucous membranes with virus-contaminated surfaces[6]. SARS-CoV-2 environmental spread through contaminated surfaces is thought to be minor as compared to direct contact, droplet transmission, or airborne transmission. [7],[8],[9].

2. Data processing and implementation

To implement the database, the data included were analysed to establish the mathematical representation of the pandemic tendency in Romania as follows:

- Analysed data to ensure no gaps between days and information. Collected missing data to complete the database.
- Processed data for each day. For instance, to acquire relevant data for Romania, information from all counties was combined and analysed. [10].

In this paper, the transfer functions derived from the structure of a Multiple-Input Single Output (MISO) mathematical representation (see in Figure 1) are identified using of input-output data pairs that have been categorized as described above[11], [12].

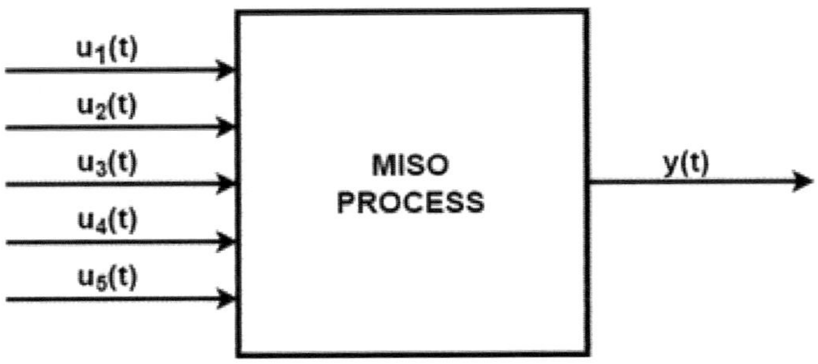

Figure 1. The MISO system

The commonly utilized technique of determining the estimated value of a non-minimal order transfer function using a collection of simplified assumptions—such as constant variance errors, known order, and delays—is widely used nowadays in techniques for establishing this MISO system. However, identifying the sequence of the transfer function that connects each input signal to each output signal is exceptionally difficult due to the complexities of the biological processes involved, particularly in the context of Sars-CoV-2 viral transmission.

For the presented system the five input signals which have been chosen are:

- Inside temperature;
- Outside temperature;
- Humidity;
- The number of tests which were taken every day;
- The quantified value of Covid-19 measures which were taken by the local authorities to control the expansion of the Covid-19 pandemic.

The transfer functions corresponding to each input signal in relation to the output signal (the number of cases/day) are determined.

IOP Conf. Series: Materials Science and Engineering **1320** (2024) 012017 doi:10.1088/1757-899X/1320/1/012017

3. Obtained results

Modelling is an important step in the parameter estimation process. Identification methods that employ input-output measurements are used in numerous situations when a detailed mathematical knowledge of the system under investigation is not required, but forecasting the system's evolution is sufficient.

In Figure 2 it can be observed that the estimated curves are following the real data curve, with some slight changes. These changes occur because the overall forecast changes when the number of cases increases or decreases.

In Figure 3, the best fit (the red curve) is presented (92.31%) together with the real data curve (the black curve).

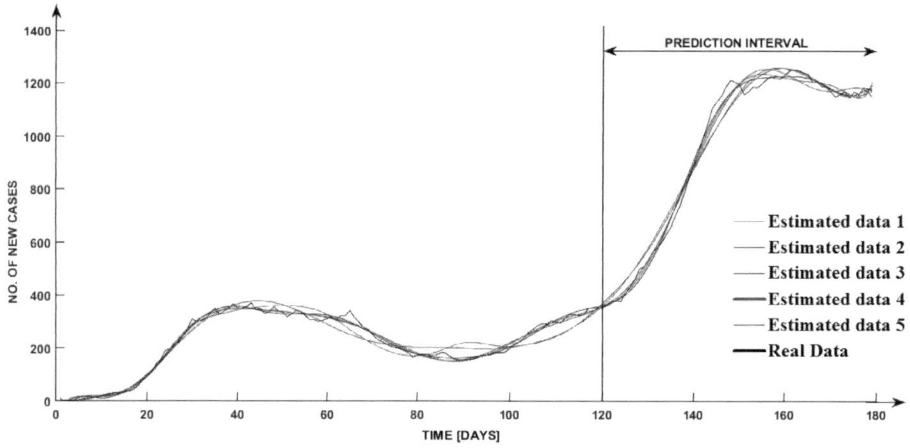

Figure 2. The representation of the transfer functions estimation versus the real data.

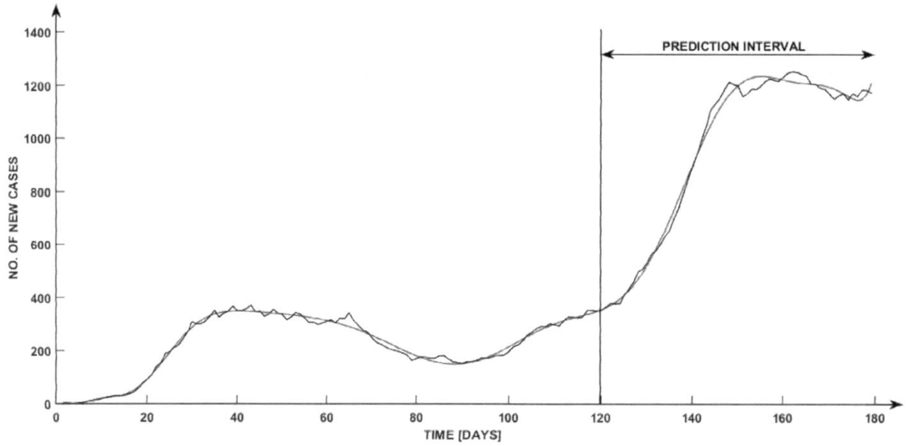

Figure 3. Predicted versus validation models.

4. Conclusions

This article identified that most conventional models and statistical analyses on COVID 19 transmission and progression to far have focused on the evolution of the COVID 19 pandemic in a single nation, as does our model, which employs a time series data from Romania.

Our study was done using publicly accessible data that did not contain personal information, thus no ethical approval was necessary.

The prediction period of the presented model is limited because, during the first 180 days of the pandemic, the examined signals (both inputs and outputs) had values extremely specific and did not span their variation on the whole domain.

It is important to accurately and specifically predict the spread of COVID-19. There is a need for evaluating viral patterns of transmission based on existing data. Most of the governments are strongly dependent on these estimations when deciding the next regulations whether it's allocating medical resources or determining whether to decrease or increase the frequency of lockdowns.

References

[1] Danesh, Farshid & Dastani, Meisam & Ghorbani, Mohammad. (2021). Retrospective and prospective approaches of coronavirus publications in the last half-century: a Latent Dirichlet allocation analysis.

[2] Karimi Shahri M, Niazkar HR, Rad F. COVID-19 and hematology findings based on the current evidences: A puzzle with many missing pieces. Int J Lab Hematol. 2021 Apr;43(2):160-168. doi: 10.1111/ijlh.13412. Epub 2020 Dec 2. PMID: 33264492; PMCID: PMC7753300.

[3] Niazkar HR, Niazkar M. COVID-19 international outbreak and the need for a suitable estimation model: a second-order polynomial equation with constant coefficients based on imported infected cases seems inadequate. Asian Pac J Trop Biomed. 2020;13:185-186. Wolters Kluwer Medknow Publications.

[4] Yaghoby, Reza. "A new nonlinear mathematical model for forecasting the waves of the COVID-19 epidemic." (2021).

[5] Gong, J, Ou, J, Qiu, X, Jie, Y, Chen, Y, Yuan, L, et al. A tool to early predict severe corona virus disease 2019 (COVID-19) : a multicenter study using the risk nomogram in Wuhan and Guangdong, China. Clin Infect Dis 2020;71:833–40.

[6] Li J, Li SL, Cai YR, et al. Epidemiological and clinical characteristics of 17 hospitalized patients with 2019 novel coronavirus infections outside Wuhan, China. MedRxiv: https://www.medrxiv.org/content/10.1101/2020.02.11.20022053v1.

[7] Zhao S., Lin Q., Ran J., Musa S.S., Yang G., Wang W. Preliminary estimation of the basic reproduction number of novel coronavirus (2019-nCoV) in China, from 2019 to 2020: a data-driven analysis in the early phase of the outbreak. Int J Infect Dis. 2020;92:214–217. doi: 10.1016/j.ijid.2020.01.050.

[8] Hossain, M. Ibrahim, Esha Binte Shahriar, Anamika Sarker, Nafisa Lubna, Mehjabeen Haque and Md. Ibrahim Khalil. "EFFECTS OF NOVEL CORONAVIRUS DISEASE-19 (COVID-19) ON VITAL ORGANS IN PATIENTS INFECTED WITH SEVERE ACUTE RESPIRATORY SYNDROME CORONAVIRUS-2 (SARS-COV-2)." Universal Journal of Pharmaceutical Research (2021).

[9] WHO Coronavirus Disease (COVID-2019) Situation Reports: https://www.who.int/emergencies/diseases/novelcoronavirus-2019/situation-reports/.

[10] Shiley KT, Nadolski G, Mickus T, et al. . Differences in the epidemiological characteristics and clinical outcomes of pandemic (H1N1) 2009 influenza, compared with seasonal influenza. Infect Control Hosp Epidemiol 2010; 31: 676–682. doi:10.1086/653204.

[11] Van den Broeck W, Gioannini C, Goncalves B, Quaggiotto M, Colizza V, et al. The GLEaMviz computational tool, a publicly available software to explore realistic epidemic spreading scenarios at the global scale. BMC Infect Dis. 2011;11:37.

[12] Vestergaard LS, Nielsen J, Krause TG, Espenhain L, Tersago K, Sierra NB, et al. Excess all-cause and influenza-attributable mortality in Europe, December 2016 to February 2017. Euro Surveill. 2017;22(14):30506. doi: 10.2807/1560-7917.ES.2017.22.14.30506.

ICEMS-BIOMED-2024 IOP Publishing
IOP Conf. Series: Materials Science and Engineering **1320** (2024) 012018 doi:10.1088/1757-899X/1320/1/012018

Dynamic Forecasting of COVID-19 Spread Using ARMAX Models

Cristina-Maria Stancioi[1*], Vlad Muresan[1], Mihail Abrudean[1], Mihai Stelian Munteanu[2], Valentin Sita[1]

[1] Automation department, Technical University of Cluj-Napoca, Cluj-Napoca, Romania
[2] Electrical Engineering and Measurements department, Technical University of Cluj-Napoca, Cluj-Napoca, Romania

*E-mail: stancioi_cristina@yahoo.com

Abstract. The COVID-19 pandemic, unprecedented in the last 100 years, struck with the massive impact it had on humanity in multiple ways, but from the medical perspective the unpredictability of the course of the disease and the wide spectrum of clinical manifestations, with extreme serious and lethal accents for a percentage of patients, was a massive reason for concern. Therefore, researchers are still trying to decipher the body's immune response and the duration of protection against a new reinfection. Immunological memory is the basis of durable protective immunity after infections or vaccinations. Its duration after infection with coronavirus 2 (SARS-CoV-2) is still under study, but more and more data are coming to complete this puzzle. Accurately foreseeing the virus's progress is critical for allocating medical resources, managing industry operations, and maintaining economic growth. To this goal, dependable and appropriate forecasting techniques based on strong mathematical models are required. This paper investigates the use of ARMAX (Autoregressive Moving Average with Exogenous Inputs) models to forecast epidemiological trends in COVID-19 outbreaks, with an emphasis on Romania.

1. Introduction

Until 2019, coronaviruses were a large family of RNA viruses, known especially for the mild respiratory infections produced by four of its members (HCoV-NL63, HCoV-229E, HCoV-OC43, HKU1) with affinity for human respiratory cells. It is estimated that 10-15% of common respiratory ailments, the so-called "colds", are produced annually by these strains[1].

On the other hand, two other new types of coronaviruses (MERS and SARS) surprised the medical community when they triggered the two brutal epidemic outbreaks in Asia (2002-2003) and the Middle East (2012) that would give doctors an idea and about the aggressive, even lethal, potential of new emerging strains[2], [3]. The lung disease caused by SARS in Asia (2002) had a mortality of 15% in contrast to the new type of coronavirus, SARS-CoV-2, which is much more contagious, but which has a calculated mortality of 1%[4].

Furthermore, there is an extensive background of using mathematical models to acquire information concerning the distribution trends in infectious illnesses and to assess the eventual influence of various methods of intervention[5]. The frequent application of mathematical models for pandemic foreseeing highlights the necessity of developing genuine representations that make up the typical distribution patterns associated with specific diseases and social environments. More detailed models are needed, however, to properly account for fluctuations in the initial phases associated with actual epidemics and get a deeper comprehension of the methodologies involved[6].

2. Data processing and implementation

ARMAX(Autoregressive-moving-average) models are commonly used in system identification since they precisely represent a wide series of everyday operations with a low level of difficulty. ARMAX models, which present the same irregular background as Kalman filters, are widely utilized in control theory for both system estimation and control design[7].

ARMAX models are members of the equation-error system family. Their outcome is determined as the sum of three regression terms: historical inputs, outputs, and white noise samples. The observed input signal, u(t), controls the corresponding correlation of a deterministic component, while an unknown blank noise, v(t), controls the stochastic part[8], [9].

ARMAX models (Figure 1) combine historical data (autoregressive and moving average components) with exogenous inputs (external variables impacting viral propagation, such as social distancing measures, vaccination rates, and public transportation data). This combination improves the model's prediction accuracy[10].

In this paper, the main input signals are represented by the inside and outside temperature, the humidity which is impacted by the season and other factors, the number of tests which were taken every day and taking into account that during the weekends, the testing activities were decreased in frequency and the measures which were taken by the authorities based on the total number of cases.

The main steps in data processing were the following:

- Missing data points have been identified and sorted out using approximation techniques such as linear interpolation or averaging neighboring values.
- Outliers were recognized and addressed using statistical approaches to avoid distortion in model estimation.
- Data were standardized to maintain consistent scales, especially for exogenous variables, allowing for greater effectiveness in model training.

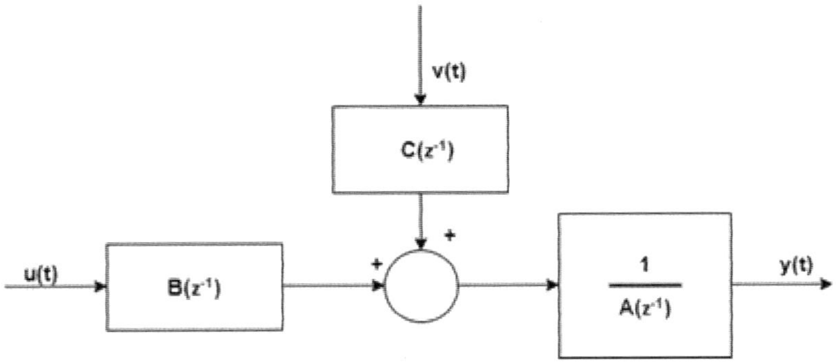

Figure 1. The ARMAX model structure

3. Obtained results

Estimating linear models using a given model structure and analyse which model structure best matches the real data model (various configurations were implemented, and the best configuration result was selected as the identification solution).

If the basic model structures do not generate helpful representations, elaborated model structures can be chosen by providing a complex model order for the similar linear model structure.

Figure 2 displays all five responses from the ARMAX models that were designed. The primary difference is a greater estimate deviation throughout the last period (the last 20 days).

Figure 3 shows the best fit of 89.42% in this example, using Estimated data 5 from the database which matches the experimental curve with satisfactory precision.

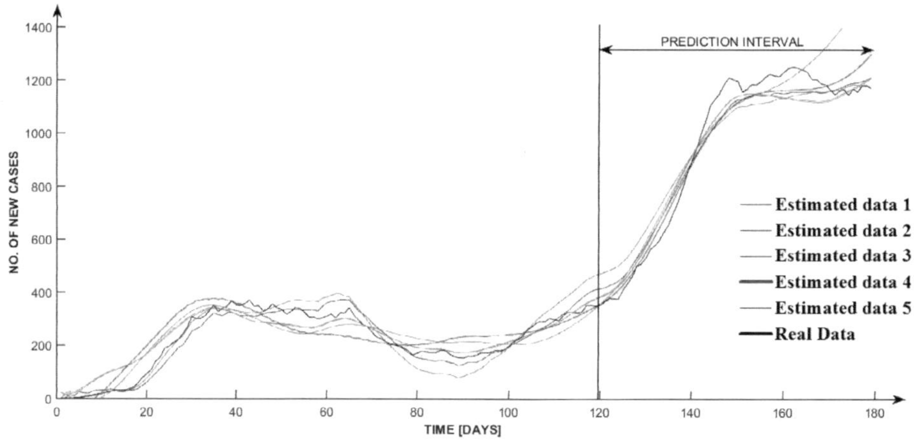

Figure 2. ARMAX-model estimated transfer functions.

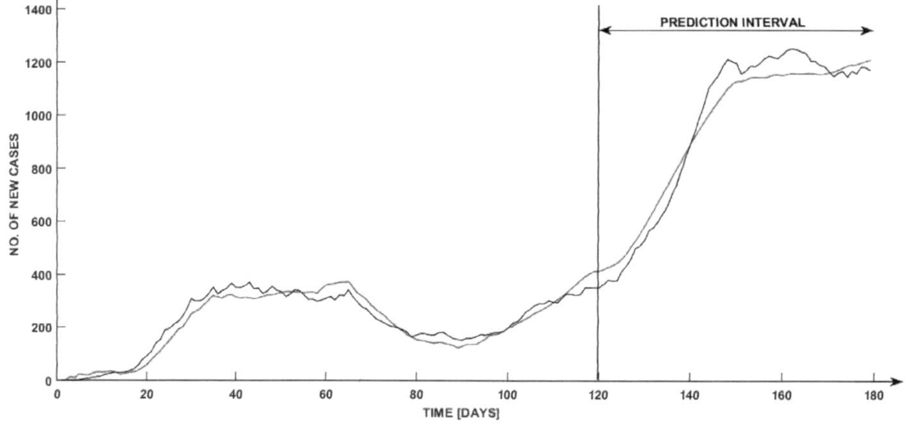

Figure 3. Estimated versus validation models.

4. Conclusions

The simulation results from the ARMAX model revealed various insights regarding the way COVID-19 spread in Romania:

1. Trend Analysis: The model effectively predicts the peaks and troughs of COVID-19 cases, demonstrating the efficacy of public health measures.
2. Resource Allocation: Predictions allows the healthcare system to plan for case spikes, assuring an adequate supply of beds, ventilators, and medical personnel.
3. Economic Impact: Forecasts guide government choices on lockdowns and restrictions, balancing public health and economic activity.

Figures on COVID-19 cases, hospitalizations, and fatalities in Romania were gathered from government databases.

Exogenous inputs, such as government intervention policies and movement trends, were also collected. The ARMAX model was designed to integrate lagged COVID-19 examples and exogenous variables, representing the dynamic link between previous values and external effects.

References

[1] Abigirl Mawonedzo (2023). Teaching practice efficacy for preservice student teachers under COVID-19: A case study of Chitungwiza schools. 10.35293/tetfle.v4i1.4272.

[2] Wang, Haidong & Wolock, Timothy & Carter, Austin & Nguyen, Grant & Kyu, Hmwe & Gakidou, Emmanuela & Hay, Simon & Mills, Edward & Trickey, Adam & Msemburi, William & Coates, Matthew & Mooney, Meghan & Fraser, Maya & Sligar, Amber & Salomon, Joshua & Larson, Heidi & Friedman, Joseph & Abajobir, Amanuel & Abate, Kalkidan & Murray, Christopher. (2016). Estimates of global, regional, and national incidence, prevalence, and mortality of HIV, 1980-2015: The Global Burden of Disease Study 2015. The Lancet HIV. 3. e361–e387. 10.1016/S2352-3018(16)30087-X.

[3] Sommariva, Silvia & Vamos, Cheryl & Mantzarlis, Alexios & Dao, Lillie & Tyson, Dinorah. (2018). Spreading the (Fake) News: Exploring Health Messages on Social Media and the Implications for Health Professionals Using a Case Study. American Journal of Health Education. 49. 1-10. 10.1080/19325037.2018.1473178.

[4] Dawood, Fatimah & Iuliano, A & Reed, Carrie & Meltzer, Martin & Shay, David & Cheng, Po-Yung & Bandaranayake, Don & Breiman, Robert & Brooks, W. Abdullah & Buchy, Philippe & Feikin, Daniel & Fowler, Karen & Gordon, Aubree & Hien, Nguyen & Horby, Peter & Huang, Sue & Katz, Mark & Krishnan, Anand & Xmf, Dtjgd & Widdowson, Marc-Alain. (2012). 12. 687-95. 10.1016/S1473-3099(12)70121-4.

[5] Santos-Lozano A, Calvo-Boyero F, López-Jiménez A, Cueto-Felgueroso C, Castillo-García A, Valenzuela PL, Arenas J, Lucia A, Martín MA; COVID-19 Hospital '12 Octubre' Clinical Biochemisty Study Group. Can routine laboratory variables predict survival in COVID-19? An artificial neural network-based approach. Clin Chem Lab Med. 2020 Oct 2;58(12):e299-e302. doi: 10.1515/cclm-2020-0730. PMID: 33001844.

[6] Merler S. Effects of clustered transmission on epidemic growth Comment on "Mathematical models to characterize early epidemic growth: A review" by Gerardo Chowell et al. Phys Life Rev. 2016 Sep;18:112-113. doi: 10.1016/j.plrev.2016.08.005. Epub 2016 Aug 12. PMID: 27545419.

[7] Davidson MW, Haim DA, Radin JM. Using networks to combine big data and traditional surveillance to improve influenza predictions. Nature 2015; 5:8154.

[8] Dembek, Zygmunt F, Tesema Chekol and Aiguo Wu. "Best practice assessment of disease modelling for infectious disease outbreaks." Epidemiology and Infection 146 (2018): 1207 - 1215.

[9] WHO Coronavirus Disease (COVID-2019) Situation Reports: https://www.who.int/emergencies/diseases/novelcoronavirus-2019/situation-reports/.

[10] Chen N, Zhou M, Dong X, et al. . Epidemiological and clinical characteristics of 99 cases of 2019 novel coronavirus pneumonia in Wuhan, China: a descriptive study. Lancet 2020; 395: 507–513. doi:10.1016/S0140-6736(20)30211-7.

IOP Conf. Series: Materials Science and Engineering **1320** (2024) 012019
doi:10.1088/1757-899X/1320/1/012019

mHealth application for cognitive rehabilitation

R Ciorap[1*], D Andrițoi[1*], M Bojoga[1,2], D Munteanu[3] and M I Ungureanu[4]

[1] Biomedical Sciences Department, Faculty of Medical Bioengineering, Grigore T. Popa University of Medicine and Pharmacy Iasi, Romania
[2] Faculty of Automatic Control and Computer Engineering, Gheorghe Asachi Technical University of Iasi
[3] Medicine I Department, Faculty of Medicine, Grigore T. Popa University of Medicine and Pharmacy Iasi, Romania
[4] Preventive Medicine and Interdisciplinary Department, Faculty of Medicine, Grigore T. Popa University of Medicine and Pharmacy Iasi, Romania

*E-mail: doru.andritoi@umfiasi.ro, radu.ciorap@umfiasi.ro

Abstract. The present work aims to develop a "serious game" application for cognitive rehabilitation after a stroke. This app (smartphone/tablet/PC) aims to provide an easily accessible and inexpensive digital tool for patients with various cognitive impairments. Using the app, patients can perform exercises structured on several difficulty levels, targeting different aspects of cognition such as memory, attention, perception of details, and language skills. These exercises are carefully designed to improve impaired cognitive functions and stimulate recovery.

1. Introduction

Cognition refers to how humans acquire, process, store, and use environmental information. Cognition is not a single concept; it includes multiple domains, including attention, executive function, visuospatial ability, memory, and language. Cognition is closely related to brain function and is essential for adaptive behavior, learning, and social interactions [1]. Impaired cognitive function can lead to decreased functional capacity, affecting rehabilitation outcomes after stroke. Decreased functional capacity (the ability to perform daily activities in a standard or acceptable manner) negatively impacts survivors' quality of life [2]. Cognitive deficits post-stroke are common and can vary in type and severity, with the areas most affected being attention and executive functions. Although cognitive impairment is as common as other neurological deficits, such as motor and sensory impairments, cognitive impairment is often overlooked in subsequent follow-up of stroke survivors unless it has progressed to dementia [3].

A transitional stage between healthy aging and dementia is represented by mild cognitive impairment (MCI), considered by researchers and clinicians as a "window" of opportunity for interventions that could delay progression to dementia. The global prevalence of dementia is expected to increase from 46.8 million in 2015 to 131.5 million in 2050 [4]. In the last few years, technology-based cognitive interventions have increased research interest worldwide. Some studies of brain training and rehabilitation for use by individuals with MCI and dementia suggest that technology holds promise to improve both cognitive and non-cognitive outcomes [5]. The studies utilized various technologies, such as computers, tablets, virtual reality (VR), and gaming consoles. General and domain-specific cognitive training have improved cognition, mainly enhancing memory, with some evidence indicating that executive function may also benefit.

Cognitive rehabilitation is restoring or compensating for cognitive deficits and related consequences through well-established therapy programs [6]. Cognitive rehabilitation aims to improve a person's ability to process and interpret information and function in all aspects of life-related to family and community [6]. Restorative cognitive rehabilitation aims to improve, strengthen, or normalize impaired cognitive functions. Restorative (direct) strategies are "exercise-like" in that they may involve intensive and repetitive use of a particular cognitive process, gradually increasing the level of difficulty or processing requirements [7]. Cognitive rehabilitation exercises are repetitive, exposing the brain to multi-sensory stimuli that help improve the plasticity of neurological functions. This process of remodeling and reorganizing brain connections helps restore and refunction the affected areas, thus allowing the recovery of impaired cognitive functions. In addition, continuous and repeated brain stimulation through cognitive exercise can enhance the brain's ability to adapt and learn, promoting a faster and more complete recovery [8].

Globally, technological advances have led to the increased accessibility of mobile phones, tablets, and computers as valuable tools in the cognitive rehabilitation process. Because of their ability to provide quick and easy access to information and communication, mobile phones, tablets, and computers can play a crucial role in the rehabilitation process. These devices allow patients to access specialized apps that support them in various aspects of cognitive recovery, such as functional readaptation, cognitive rehabilitation exercises, and autonomously managing daily activities [9]. Integrating technology into the rehabilitation process can create effective continuity between therapy sessions and the recovery period at home. Through apps, patients have exercises, instant feedback, and progress monitoring. Technology can also boost patients' motivation and engagement, giving them a sense of control and active involvement in their cognitive recovery [9]. In [10], they concluded that simple, interactive, and easy-to-use features of mHealth-based assistive technology ensure better healthcare services for people with cognitive impairments.

2. Materials and Methods

The application presented in this paper offers a convenient, modern, and interactive alternative to traditional paper-and-pencil exercises, potentially boosting patients' motivation and involvement in the cognitive rehabilitation process. The app can be used independently by patients at home and in therapy sessions under the supervision of a specialist. The cognitive rehabilitation process needs to be intensive and long-term, which is why the development of this app aims to provide a digitized way of continuing cognitive rehabilitation outside therapy sessions, thus improving patients' autonomy and quality of life.

A serious game app has been developed that is compatible with various devices such as PCs, tablets, and smartphones. This app is designed to help stroke patients in their cognitive rehabilitation process. The app has several functionalities intended to support the recovery and restoration of patients' mental abilities.

The app was developed on the Unity platform created by Unity Technologies, providing an integrated game development environment. Unity uses a system known as "component-based development, " making it easy to build and organize game or app elements through reusable components. The platform offers a wide range of tools and functionality to support the development of games and interactive applications; these include a powerful editor that allows the creation and modification of scenes, characters, objects, and game logic, a realistic physics system, an advanced graphics engine, support for 2D and 3D animations, resource management and more.

When the patient opens the application, they must fill in personal information such as first name, last name, and age in the fields provided. Once the information is entered correctly, the patient can proceed to the next step.

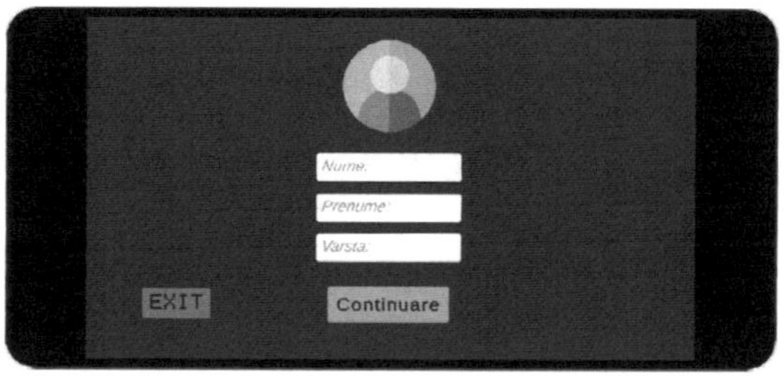

Figure 1. User interface for data input

Once the patient has entered their details, they can choose a difficulty level from those available: easy, medium, or hard. The patient will also be able to select a theme for the exercise they want to do, such as colors or animals.

Figure 2. Interface for selecting difficulty level and theme

Once patients choose the difficulty level and topic, they can select one of the three available exercises. The first exercise consists of finding pairs of cards. The cards can be of two types: color-themed or animal-themed (offering diversity and more stimuli). The difficulty increases by increasing the number of cards available: from 8 cards (4 pairs) to 12 cards (6 pairs) and finally to 18 cards (9 pairs). This exercise can help improve visual and working memory, as the patient has to memorize the position of the cards and remember where the pairs are to find them in as few moves as possible. Concentration and attention are also stimulated, as the patient has to pay attention to the colors or pictures on the cards and associate them correctly. Last but not least, depending on the number of pairs to be found, the patient has to plan the moves and organize the cards in memory to maximize efficiency and find the correct pairs as quickly as possible.

The number of moves made (incremented each time a card is turned face up), the time from turning over the first card to completing the exercise, and the score are recorded. For each correct pair made, the score increases by 4 points. If a card has been previously turned over and no pair has been found, 1 point is deducted from the score. The maximum scores for this exercise are 16 for the low difficulty level, 24 for the medium difficulty level, and 36 for the difficult level.

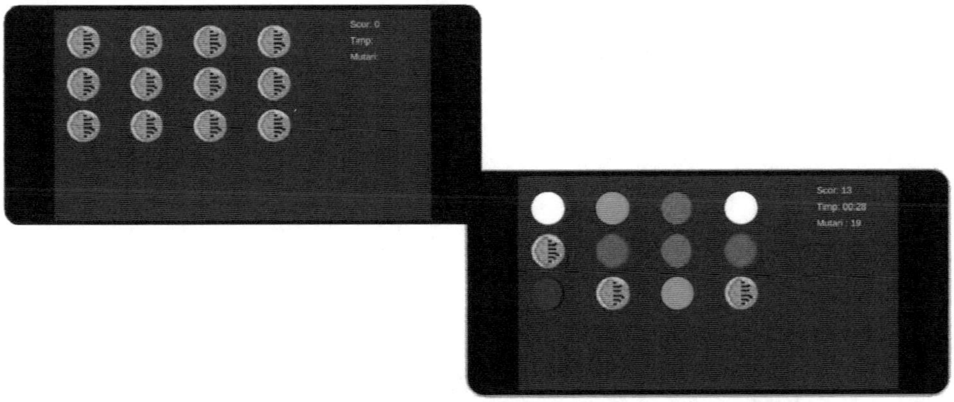

Figure 3. Card pairing exercise interface - medium difficulty: ***A***. All cards are face down; ***B***. The user has started pairing

The second exercise involves recognizing shapes and colors. Depending on the difficulty level selected, the patient will be asked to identify different shapes and colors: 10, 15, or 20. The patient will be presented with a specific shape and color object. In one box, they will have available shape options to choose from by ticking a checkbox (expressed by words, such as circle, square, etc.), and in a second box, they will have available color options to choose from by ticking a checkbox (expressed by words, such as white, red, etc.). After selecting the answers that the patient considers to be correct, they can check them by pressing the "Check" button. Suppose they have chosen both answers correctly. In that case, the message "Correct" will be displayed in each box, and if they have selected one of the answers incorrectly, the message "Wrong" will be displayed in the box corresponding to the answer. If both answers were selected wrong, "Wrong" will be displayed in each box. After checking the answers, the "Check" button changes to "Continue". Pressing it will display the next item.

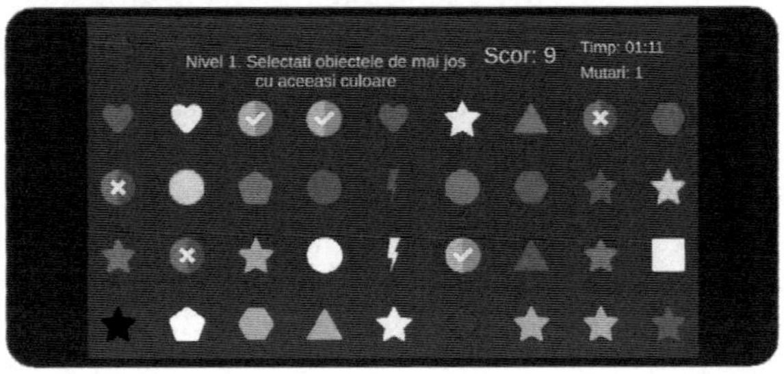

Figure 4. Shape or colour matching exercise - low difficulty: correct answers are marked by a green circular shape with a positive symbol and wrong answers by a red circular shape with a negative symbol

This exercise targets several cognitive functions, including visual perception, language skills, attention, and concentration.

The third exercise involves the process of matching shapes and colors. The difficulty level determines the patient's specific requirement, which may consist of associating the color, shape, or both characteristics of an object shown in the top left corner of the screen with other objects arranged in a matrix on the screen.

Thirty-six objects of different shapes and colors are displayed on the screen, and the object to be matched is in the top left corner. This exercise is structured in four rounds, i.e., in each session, the patient will associate the color/shape/ambles (depending on the difficulty) with new objects generated in each round. For each correct association, the object in the matrix forms a green circle, indicating the validation of the answer. For each wrong association, the object forms a red circle, indicating the error of the answer.

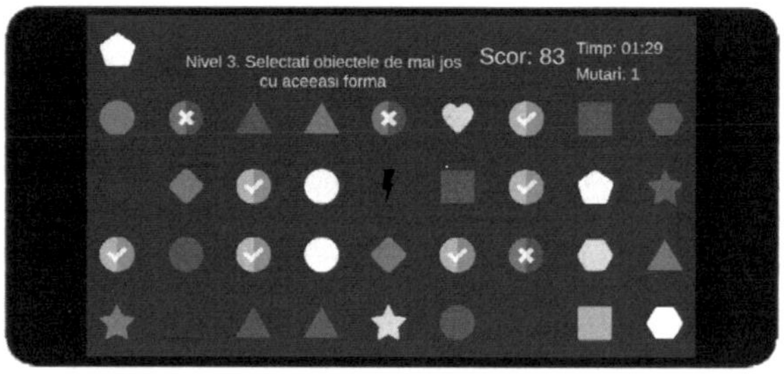

Figure 5. Shape/colour matching exercise interface - medium difficulty: shape matching; top right corner shows score, time and number of moves; green circles indicate correct answers, red circles indicate wrong answers

After each exercise, the patient is given a feedback page with relevant information, such as the score obtained and the time the training was completed. The patient has various options on this page to manage the following actions within the application. These options include exiting the app, restarting the previous exercise, accessing the main menu (difficulty and theme choice), or accessing the personal data entry section if a therapist uses the same device to work with several patients.

3. Discussion and conclusions

Progress monitoring can provide patients with concrete and measurable feedback on their progress. The results can boost motivation and commitment to rehabilitation, encouraging patients to go beyond their limits and achieve their goals. By being aware of their results and progress, patients can develop a positive attitude and confidence in their ability to recover.

The "Animals" theme is only available for the card pairing exercise. To avoid confusion and provide a more coherent experience, we are considering adding the theme "Animals" to the other exercises available in the app. In addition, more difficulty levels will be introduced to provide a wider range of challenges and allow patients to adapt to the difficulty level according to their progress. It will ensure a personalized experience tailored to each patient's abilities and needs within their cognitive rehabilitation program.

The app is easily accessible and usable on various platforms (phone, tablet, PC). It allows patients to perform exercises in cognitive rehabilitation games at any time and place, thus allowing them to engage in an intensive and continuous recovery program. Patients are provided with three games that address multiple cognitive functions, such as attention, memory, perception, logical thinking, and language, helping to train and stimulate different aspects of impaired cognition.

Developing an app for post-stroke cognitive rehabilitation is not easy and quick. It requires considerable effort and many factors to consider. The depth and complexity of the condition must be understood for creating an app. Therefore, extensive research is necessary to comprehend cognitive rehabilitation's theoretical and practical basis after stroke and identify patients' needs and requirements.

In conclusion, developing a "serious game" application for post-stroke cognitive rehabilitation is a promising approach capable of overcoming the limitations of traditional techniques. However, it is essential to note that developing such an application is a long-term endeavor, requiring multidisciplinary knowledge and a complex approach. Still, it can bring significant benefits to patients.

References

[1] Cumming T B, Marshall R S and Lazar R M 2013 *Int J Stroke* **8** (1), 38-45.
[2] Fitri F, Fithrie A and Rambe A 2020 *International journal of Research Science and Management* **7**.
[3] Kalaria R N, Akinyemi R and Ihara M 2016 *Bba-Mol Basis Dis* **1862** (5), 915-925.
[4] Anderson N D 2019 *CNS Spectrums* **24** (1), 78-87.
[5] Ge S, Zhu Z, Wu B and McConnell E S 2018 *Bmc Geriatr* **18** (1), 213.
[6] Nejati V, 2023, Academic Press.
[7] Koehler R, Wilhelm E E and Shoulson I, 2011, *Cognitive Rehabilitation Therapy for Traumatic Brain Injury: Evaluating the Evidence*, THE NATIONAL ACADEMIES PRESS.
[8] Nie P Y, Liu F, Lin S H, Guo J Y, Chen X M, Chen S L, Yu L Q and Lin R H 2022 *J Clin Nurs* **31** (9-10), 1136-1148.
[9] Marwaa M N, Guidetti S, Ytterberg C and Kristensen H K 2022 *J Rehabil Med* **54**, jrm00269.
[10] Yousaf K, Mehmood Z, Awan I A, Saba T, Alharbey R, Qadah T and Alrige M A 2020 *Health Care Management Science* **23** (2), 287-309.

Modulation influence on power deposition inside a human head model exposed to 3.5 GHz plane wave

Vivian Mihaela Maftei[1*], Remus Stanca[2] and Simona Miclaus[1]

[1] "Nicolae Balcescu" Land Forces Academy, Sibiu, Romania
[2] "Lucian Blaga" University, Sibiu, Romania

[*] maftei_vivi@yahoo.com

Abstract. Biological effects of radiofrequency fields remain one major subject of interest, with the continuous development of communication technologies and their more and more intensive use. Many authors have discussed the special role that modulation has on the produced effect, comparatively to continuous wave. Numerical dosimetric studies following the differences in power deposition in tissues, in function of the applied waveform, are scarce. Here, we analyse by numerical dosimetry, the specific absorption rate (SAR) of energy absorbed in a human head model when a plane wave, either a continuous one or two modulated ones (amplitude-modulated and quadrature amplitude modulated) are impinging the head, from frontal or lateral directions. The final objective was to emphasize the differences, based on average, peak values, or tissue-specific values of SAR. In all investigated cases the carrier frequency was 3.5 GHz, which belongs to the 5G mobile communications in the FR1 band. The computations were made in the case of the same total incident energy delivered per each modulation considered.

1. Introduction and objectives

Human exposure to electromagnetic fields (EMF) continues to represent a significant interest in research, mainly with the recent introduction of the 5th generation of mobile communication technologies (5G). This interest is due either to the increase in frequency (which is equivalent to higher intrinsic energies) up to millimetric waves, or to the lack of knowledge on the impact of very high and rapid changes in the momentary incident powers upon biological responses.

Continuous debates on safety exposure standards proofs regarding thermal versus non-thermal or athermal biological effects, require continuation of both dosimetric studies and biological analyses [1]. From this perspective, frequencies higher than 6 GHz and signals characterized by high crest factors, should receive special attention.

Interesting to note is the fact that, even well documented to date, from the biological database point of view, modulation type and its parameters effect on power deposition in tissues are rarely investigated from the point of view of dosimetric studies. Some experimental approaches have been done, including frequencies belonging to 5G communications [2], [3]. However, theoretical dosimetry is very scarce, as far as we know [4].

The newest modulation technologies incur risks, prior to release into the marketplace [5]. Short-term impact such as insomnia, memory, reaction time, hearing and vision, should be carefully investigated. Modulations, doses and dose rates, peak exposures, may be of particular interest over the long term, as well. Authors of [6], in a comprehensive review finalised in 2011 on the biological effects of modulated radiofrequency (RF) fields showed that even if the majority of the studies have reported no modulation-specific effects, a few interesting exceptions existed, and they indicated that there may be specific effects from amplitude modulated (AM) signals, mainly on the human central nervous system.

In a recent paper from 2022, Lai and Levitt [7] deeply analyse and discuss the exceptional role that intensity, exposure duration, and modulation have on the triggered biological effects of RF radiation and exposure guidelines. The authors brought proofs that electromagnetic energy is more biologically potent in causing effects than thermal changes (thru dielectric heating). The two energies act through different mechanisms, and the first one is presently the challenging one. Moreover, the research database cumulated until now can be separated, based on the four basic unpredictable and contradictory research outcomes that occurred with continuous wave (CW) and modulation. The general classification shows that [7]:

a) CW and modulation produce different effects.
b) Different forms of modulation produce different effects.
c) Modulation produces effects, but not CW.
d) Modulation and CW have different biological effectiveness/potencies.

Two papers were found in the literature to treat dosimetrically the differences in power deposition when using CW versus modulated signals.

Busygina et al. [8] investigated the SAR distribution differences in three biological tissues (blood, fat and muscle) using three types of incident waveforms at 900 MHz: Gaussian pulse (GP), bipolar-Gaussian pulse (bGP) and AM signal. A transverse electromagnetic cell was used in both the simulation and experimental measurements. The results showed differences in SAR values, depending both on the applied waveform and on the tissue dieletric properties. The highest SAR values were generally obtained when the AM signal was applied.

The authors of [9] conducted a numerical simulation study to analyse SAR variation inside a water spherical phantom of 25 cm diameter, when either a square-wave modulated signal at 1GHz, or a CW was applied as the stimulus. They concluded that both exposures were substantially equivalent if the CW signal carried the same average power as the modulated one.

Present paper aims to compute the local and average SAR values inside a human head model irradiated by a plane wave but with three different waveforms, from two different directions, in order to emphasize the differences in power deposition at 3.5 GHz. The comparisons are made for the case when the incident wave has the same incident energy for all three modulation types and the same maximum reached amplitude during the transmission duration. The final objective is the theoretical study of modulation influence on the human electromagnetic exposure at the frequency used by a 5G-FR1 mobile phone.

2. Human head model, delivered waveforms and dosimetric computations

A realistic head model composed of 16 tissues (Fig. 1a) was imported in CST Studio Suite and was used for the dosimetric computations in the time domain. A plane wave was exciting the head either frontal or lateral (Fig. 1b), this choice being applied because the use of a mobile phone can take place in these configurations. The central frequency was 3.5 GHz, while three waveforms were successively used in the computations for their absorption in the head (Fig. 2): a) CW in the form of a GP with the duration of 71.2 ns (implicit GP) and maximum amplitude of 1V; b) AM signal with the duration of 21.65 ns and the same maximum amplitude; c) 64-quadrature modulation (64-QAM) with the duration of 15.5 ns and the same maximum amplitude. The

IOP Conf. Series: Materials Science and Engineering **1320** (2024) 012020 doi:10.1088/1757-899X/1320/1/012020

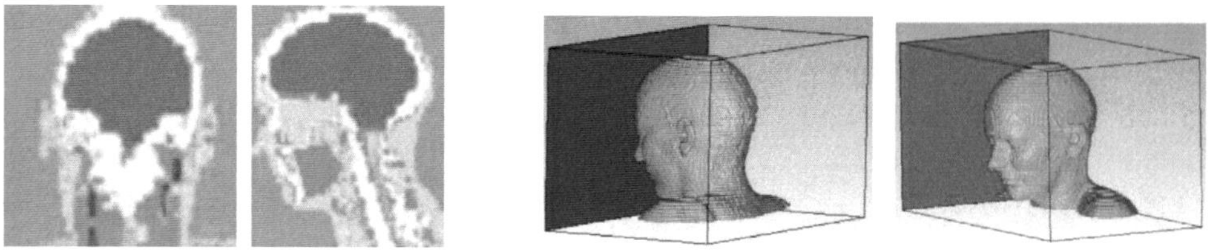

(a) (b)

Figure 1. a) The human head dielectric and geometric model (back and lateral views); b) The human head model impinged frontal (left) or lateral (right) by the plane wave stimulus.

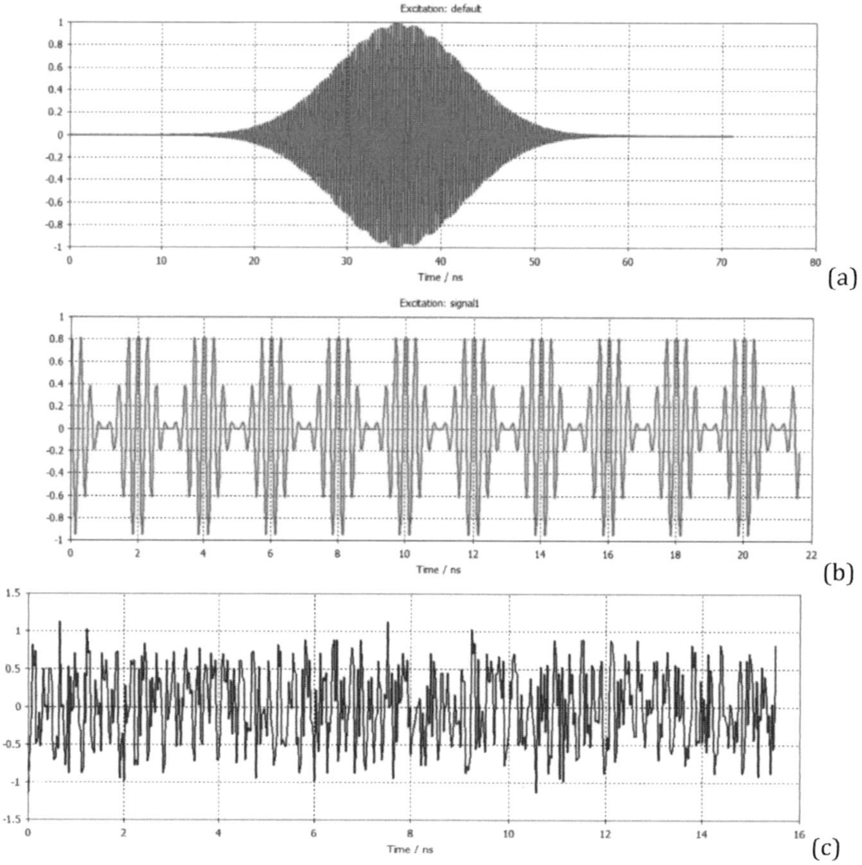

Figure 2. Stimuli applied for dosimetric computations (incident signal types): a) Gaussian pulse (continuous wave – CW); b) Amplitude modulation (AM); c) Quadrature amplitude modulation (64-QAM).

durations of the AM and 64-QAM signals were calculated by equalizing the total incident energies of the three applied waveforms (GP was considered as a reference). With this imposed condition, correct comparisons ca be made regarding SAR values. SAR computations were obtained by both the usual SAR post-processing (for average and peak vales) and by power loss density monitor (for tissue-specific SAR values). The mesh used for discretization of the calculation space was non-uniform and hexahedral, the CST software using an adaptive mesh refinement to reduce the computational error.

3. Results and discussion

Section planes were made in the head model to observe the local SAR distribution. The SAR averaged over 1g (SAR_avg_1g) of tissue is represented in the maps. Fig. 3 presents the three modulation excitation effects as the figure lines defined by a, b, c. The two columns of Fig. 3 refer to the two exposure directions – either frontal or lateral. It is observed that in the central section plane represented, all the SAR value distributions are different and CW wave produces much larger (roughly two orders of magnitude) SAR values than the other two modulations. These situations are due to the very different values of the "accepted power" (which is the input power

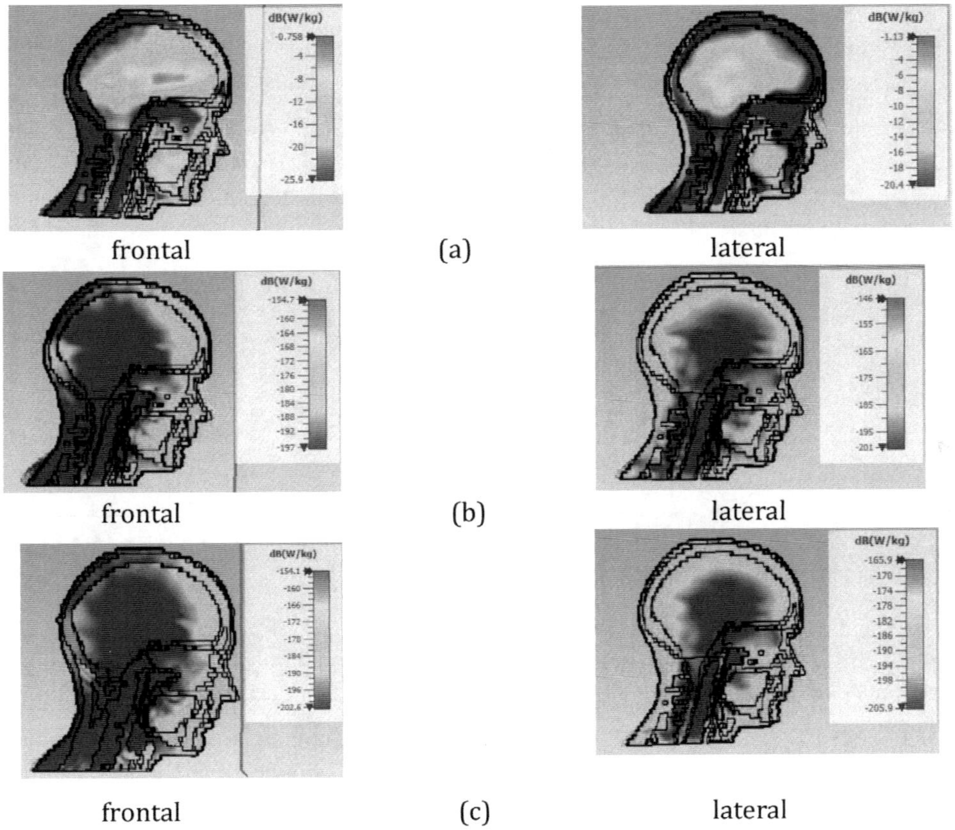

Figure 3. SAR_avg_1g distribution inside the head (sagittal section) for both incidence directions in case of : a) CW exposure; b) AM exposure; c) 64-QAM exposure.

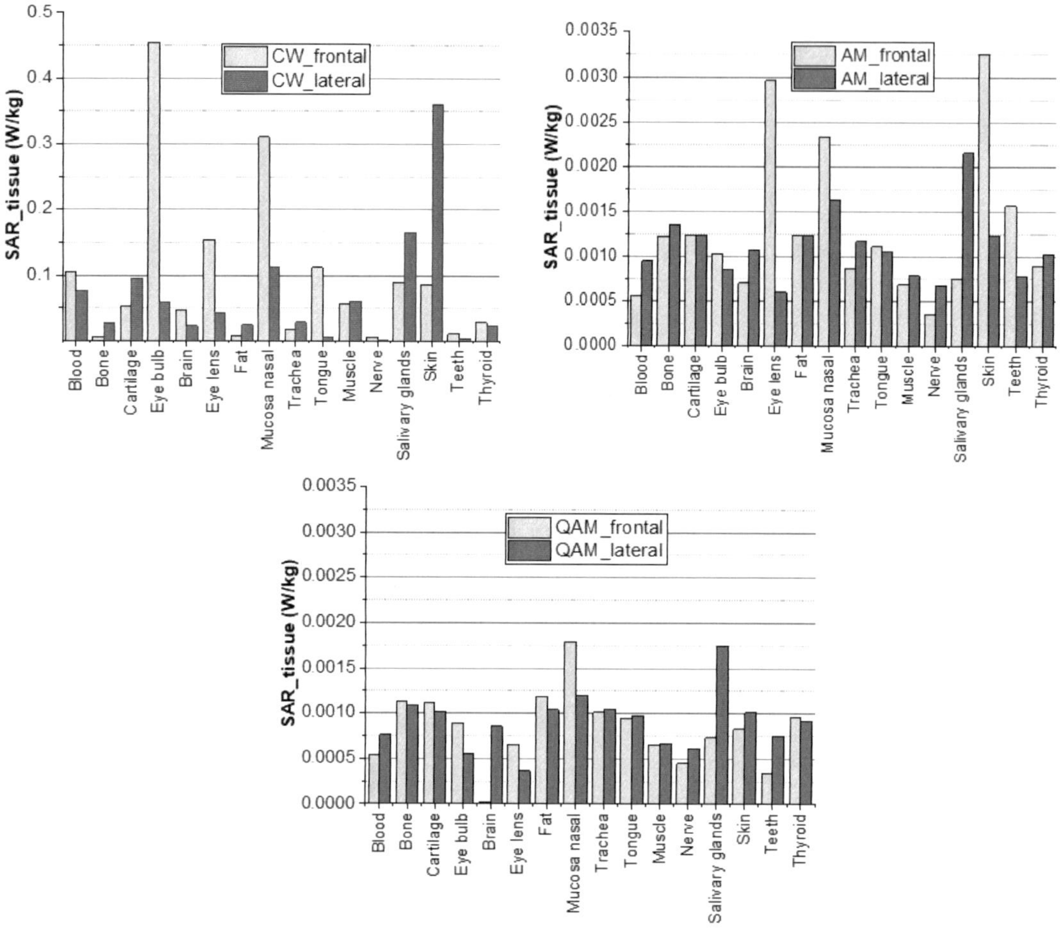

Figure 4. SAR values calculated in each tissue in the head for the three types of modulations applied and for both incidence directions of the waves

of the SAR calculation) in the three exposures. Practically, the three "accepted power" values were: a) CW – 0.5W; b) AM – 1.3 x 10^{-17}W; c) 64-QAM – 3.3 x 10^{-18}W. Such very different values of the power influenced the results so that the comparison of absolute values in the images of Fig. 3 is not possible immediately. The same effect is observable in Fig. 4 where we represented the SAR vale in each tissue. The comparison between the three graphs is again not possible for Fig. 4, but the comparison of power deposition between the tissues inside each of the three modulation cases can be successfully done.

Fig. 4 clearly shows how different SAR values between tissues are, and how the direction of the wave incidence is also influencing the values (green versus grey bars). Interesting to note is the fact that the highest differences between tissues and wave orientation happened for the CW exposure of the head (upper left graph in Fig. 4). In the CW case, eye bulb, mucosa nasal and eye lens are the most affected tissues for the frontal exposure to the field. If CW wave is incident from lateral to the head, the most affected tissues are the skin, salivary glands and mucosa nasal. Overall, eyebulb and skin are mostly affected by the CW at 3.5 GHz. Also, in CW case the SAR values of the tissues are very spread, which is not the case when modulated signals are applied. The most

IOP Conf. Series: Materials Science and Engineering **1320** (2024) 012020 doi:10.1088/1757-899X/1320/1/012020

similar/uniform values of SAR distribution among tissues is observed for the 64-QAM modulation. Again, we should remember that the absolute values of SAR in the three cases of modulation cannot be compared. In the case of AM signal, the frontal exposure mostly affects the eye lens, mucosa nasal and skin, while for the lateral exposure the salivary gland, mucosa nasal and bone are the most affected. Generally, interesting differences are noted between power loss distributions in the tissues – very variable for CW and much more uniform for QAM signal.

In order to compare the absolute values of SAR among modulations, we had to normalize everything to the « accepted power ». In Fig. 5 we represented this normalization applied to the averaged SAR in the head over any 1g of tissue, over maximum SAR and over maximum point SAR values. In all three situations, the lateral exposure created higher power absorption than the frontal exposure. The highest effect is due, in order, to AM modulation, followed by 64-QAM modulation and finally the lowest impact is due to CW.

Another interesting characteristic is that AM and 64-QAM modulations create large differences in SAR values (average, maximum and point peak), in function of the incidence direction of the wave.

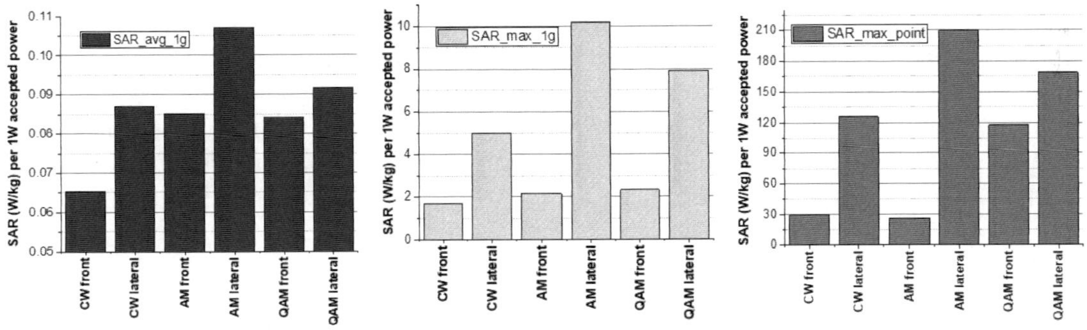

Figure 5. Normalized SAR values calculated per 1W accepted power, as average values over 1g (left), as peak value (center) and as maximum in a single point in space (right)

The largest ratio between extreme values of average SAR over all modulations and exposure directions is as big as 1.6 times. The largest ratio between extreme values of maximum SAR over all modulations and exposure directions is as large as 5.6 times. The largest ratio between extreme values of maximum point SAR over all modulations and exposure directions is as large as 7 times. These facts prove that modulation is one of the main factors that influences a biological effect and that a modulation like 64-QAM which is specific for 5G communications, can be more dangerous than a CW.

AM modulation may induce roughly 100 times higher average SAR values than maximum SAR, and 2000 times higher maximum point SAR than average SAR. For the 64-QAM the similar factors are 100 and 1650 respectively. Therefore, it becomes possible that hot spots generated inside the head to be more intense for modulated waves than for continuous wave.

4. Conclusions

Even if the simulation of power deposition in tissues of a human head model didn't take into account the blood circulation and the effect of heat exchange with the environment, present computations indicate that clear and significant differences may be determined by different modulations applied to the model.

SAR values obtained by simulation in CST Studio thru post-processing do not allow a correct comparison of absolute values of SAR between different incident applied waveforms. Therefore, power loss density is a good option to have access to the information regarding the stimulated power, which is always different for different modulations, even if the maximum amplitude of the signals is equal.

Current approach used two key elements in solving the correct comparison of SAR vales between different waveforms: a) the applied signals had to transport the same amount of incident electromagnetic energy; in this regard, duration of each modulation was calculated at the beginning, accordingly; b) the normalization of SAR values obtained from power loss density computations was made by dividing SAR to the "stimulated power".

All the results obtained here sustain the final conclusion that modulation also counts for different dissipation of electromagnetic power than continuous wave. In the exemplified cases of present approach, and at 3.5 GHz, the deposition effect was the highest, in descending order, for: amplitude modulation, 64-QAM modulation and continuous wave, respectively. Moreover, significant differences evolved, when we analysed separately each tissue. At this frequency, the most affected tissues overall, were: eye bulb, mucosa nasal, eye lens, skin and salivary glands. When applying the plane wave from two different directions, significant differences emerged regarding SAR values and sensitive tissue types.

References

[1] D. J. Panagopoulos et al., "Human-made electromagnetic fields: Ion forced-oscillation and voltage-gated ion channel dysfunction, oxidative stress and DNA damage (Review)," Int J Oncol, vol. 59, no. 5, p. 92, Nov. 2021. doi: 10.3892/ijo.2021.5272.

[2] R. N. Kostoff et al., "Adverse health effects of 5G mobile networking technology under real-life conditions," Toxicol Lett, vol. 323, pp. 35-40, May 2020. doi: 10.1016/j.toxlet.2020.01.020.

[3] A. Joushomme et al., "Effects of 5G-modulated 3.5 GHz radiofrequency field exposures on HSF1, RAS, ERK, and PML activation in live fibroblasts and keratinocytes cells," Sci Rep, vol. 13, p. 8305, 2023. doi: 10.1038/s41598-023-35397-w.

[4] K. Foster et al., "Transient Thermal Responses of Skin to Pulsed Millimeter Waves," IEEE Access, vol. 8, pp. 1-1, 2020. doi: 10.1109/ACCESS.2020.3008322.

[5] A. B. Miller et al., "Risks to Health and Well-Being From Radio-Frequency Radiation Emitted by Cell Phones and Other Wireless Devices," Front. Public Health, vol. 7, p. 223, 2019. doi: 10.3389/fpubh.2019.00223.

[6] J. Juutilainen et al., "Review of possible modulation-dependent biological effects of radiofrequency fields," Bioelectromagnetics, vol. 32, no. 7, pp. 511-534, Oct. 2011. doi: 10.1002/bem.20652.

[7] H. Lai and B. B. Levitt, "The roles of intensity, exposure duration, and modulation on the biological effects of radiofrequency radiation and exposure guidelines," Electromagn Biol Med, vol. 41, no. 2, pp. 230-255, Apr. 2022. doi: 10.1080/15368378.2022.2065683.

[8] A. V. Busygina and M. E. Komnatnov, "Simulating the impact of electromagnetic radiation on living tissues in normal and pathological conditions," presented at the IEEE Siberian Conference on Control and Communications (SIBIRCON), 2022, pp. 500-503. doi: 10.1109/SIBIRCON56155.2022.10017087.

[9] D. Brizi et al., "On the Specific Absorption Rate Behavior of Square-wave Modulated Signals Exposures," presented at the IEEE International Symposium on Antennas and Propagation and North American Radio Science Meeting, 2020, pp. 1559-1560. doi: 10.1109/IEEECONF35879.2020.9330103.

IOP Conf. Series: Materials Science and Engineering **1320** (2024) 012021 doi:10.1088/1757-899X/1320/1/012021

Effect of the electric field on human cells: from *in vitro* response to *in vivo* feedback

Diana V. Portan[1*]

[1] Department of Mechanical Eng. & Aeronautics, University of Patras, Patra 26504, Greece

*E-mail: portan@upatras.gr

Abstract. Experimental observations of human cells feedback to an electric field are presented, followed by a debate based on literature sources on implications of the electromagnetic fields generated by modern devices on the human health. While focus is usually on electromagnetic fields negative effects, the importance of having an appropriate bioelectric environment for the body and its crucial role in the *in vitro* studies are neglected. To enable progress, the merging of laboratory and clinical investigations is the solution to finding prevention, protection and treatment methods against the negative effects produced by electromagnetic fields.

1. Introduction

The living organisms on the Earth are electrical bodies made up of charged particles and consequently subjected to the laws of electromagnetism. The human body generates electricity that flows through it [1], while the heart produces an electric current that runs through the body and further on, to every cell. This electric current in the body generates a biologic electromagnetic field.

However, electromagnetic fields exposure caused by sources emitted from external devices enables different processes in the human body, than the inner electromagnetic ones. Changes in social trends accompanied by the need of fast and efficient communication determined several environmental changes especially in the cities. During the 20h century environmental exposure to artificial electromagnetic fields has been steadily increasing as growing electricity demand. The World Health Organization declares that body heating is the main effect of radiofrequency electromagnetic fields [2]. The immediate effect of body heating is tiredness, weakness, feeling lightheaded, and nauseated, and having a headache. Power, electromagnetic energy frequency and distance are the parameters that will produce shorter- or long-term effects on the body. Despite that, the extent to which an electromagnetic field affects the human body differs for individuals and thresholds are not clear. Several studies are contradictory and there is no classification that can precisely indicate how a certain organ or tissue will respond to an electromagnetic field of a specific intensity depending on the patient category, such as age, gender, weight, or pathologic history.

Conflicting data are provided in literature, related to experimental studies on different frequencies of electromagnetic fields, under specific *in vivo* and *in vitro* environments. The degradation of the DNA molecule has been several times reported; on the other hand, other studies present a low risk for the same magnitude of the exposure. Differences could result from the biological materials, experimental design, or devices used to create electromagnetic fields, among other reasons [3]. On the other hand, the innovative powered biomedical devices are nowadays indispensable. From smart watches to peacemakers and deep brain neurostimulators [4], the biomedical devices that function using electromagnetic fields play a central role in the transformation of the healthcare sector and for the improvement of patient safety, comfort, and lifestyle. The last generation of implantable electronics allow the connection of the peripheral nervous system with prostheses or exoskeletons [5] and therefore enable the rehabilitation of amputees and paralysed persons offering them a normal living that also involves a regular job. With all these, the only choice is to find adaptation solutions that will help the human body integrate with the new environmental challenges. In this direction we need complex *in vitro* studies accompanied by accurate clinical observations. Finally, standardization of laboratory trials and direct correlation with the *in vivo* results are demanded. The presence of an electromagnetic field can be both favourable or harmful, depending on the perspective and the context. For some's health essential, for others detrimental, the issue of the environmental electromagnetic field is still a challenge. To establish criteria and thresholds, we need to holistically observe its consequences on the biological unit, that is the cell, followed by organs and organisms, which is the scope of the present work.

2. Laboratory vs. clinical findings

2.1 Laboratory investigations

Bioelectricity in the human organism was observed since the 80s, when it has been measured. It was then found that the electrical potentials of the human body range between 10 and 60 mV at various locations. Within this limit, electric fields were considered non-harmful and even compulsory for the human cells' wellbeing. It was further proven that various biological systems are dependent on an electric stimulation having an exogenous or endogenous source. *In vitro* studies further suggested that electric fields that have a physiologic intensity are necessary to control and to adjust cellular and tissue homeostasis [6].

More recent investigations have shown that the electric field is a natural healing tool for the human skin. The healing process involves bioelectricity that is generated upon injury at the outer surface of the epidermis. It was measured that the field beneath the stratum corneum was around 150–200 mV/mm for 3 days after injury was caused. This was followed by a gradual decrease of the electricity as days passed, until it completely disappeared upon healing of the injury [7]. After several such studies were performed, it was considered that the reproduction of a more body-like environment *in vitro* will benefit the cell cultures. In this context, Portan et al [8] have previously investigated the response of bone cells to a combined effect of substrate type and an electric field. Cell cultures are extremely susceptible to micro-environmental changes. The physiologic bioelectricity in the human body enables the cells and tissues to work to their full potential. *In vitro*, bioelectricity is neglected since there are no devices to maintain the cell culture under all conditions, that is temperature, oxygen, and bioelectricity. This means that all observations of cells populations behaviour in the absence of an electric field are lacking important feedback. Home made devices that provide an electric field are used in some laboratories. The results of one such experiment are presented below.

As observed in in Fig.1a, several experimental parameters were applied to observe the feedback of human bone cells to an electric field. When exposed to harmful radiation, living cells will gradually initiate a biologic deactivation program meant to protect the population. This phenomenon is very well known in case of plant and animal organisms; if the environment is found inappropriate, organism refuse to produce offsprings. *In vitro*, at cell population level, it is expected that the reaction will be: (1) multiplication stop– necrosis in case of long-term exposure (2) dysfunctionality of any type – lack of metabolism and necrosis after moderate term exposure and (3) necrosis/ apoptosis – instant death. As observed in Fig.1a, the conditions in *Experiment 1* determined the deactivation of cells division process. The next two experiments determined cells necrosis while for *Experiment 4*, conditions were found appropriate and moreover, enabled the adaptation of the bone cells to a specific substrate which was not possible in the absence of an electric field. This has proven that an inadequate electric field is damaging but also the presence of an optimum intensity is compulsory for the correct functioning of the cells *in vitro*.

Experiment 1 1 Volt for about 20 min	**Cells gathered in groups. Cells morphology was unusual. Cells lived but did not multiply.**
Experiment 2 1 Day osteoblasts on plastic, 3 hours electrical stimulation at 1 Volt	**Necrosis**
Experiment 3 1 Day osteoblasts on plastic, 2 hours electrical stimulation at 0.7 Volts	**Necrosis**
Experiment 4 1 Day osteoblasts on a material, 1 hour electrical stimulation at 0.3 Volts	Cells survival

(a)

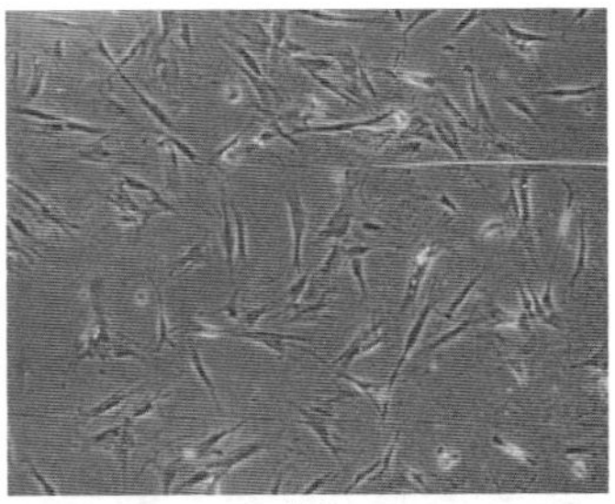

(b)

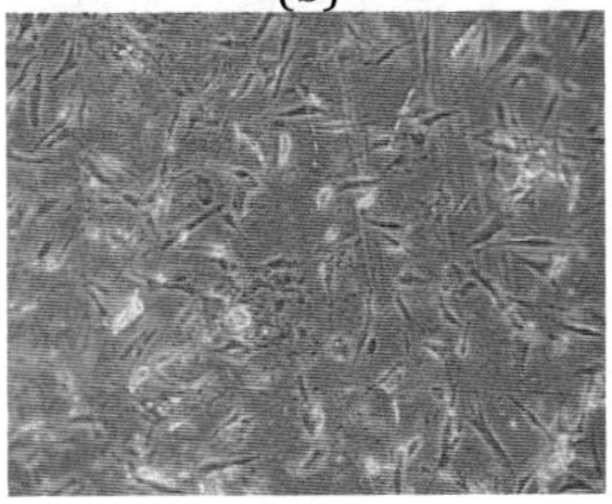

(c)

Figure 1. (a) Conditions of electrochemical stimulation of bone cells applied in a study by Portan et al [8], (b) Cells in the absence of an electric field and (c) Electrically stimulated cells

In further investigations, Mesenchymal Stem Cell type Wharton's Jelly (WJ-MSCs) were obtained as described in [9, 10] and further deposited in electrically conductive scaffolds (carbon nanotubes reinforced polylactic acid) that were fabricated by 3D printing and were exposed to the electrical stimulation (100 mV) for 2 hours/ day, for the first 3 consecutive days of incubation. The results showed an improved level of osteopontin in electrically stimulated cells as well as a considerable improvement in the total protein level. The other biomarker values (osteocalcin and alkaline phosphatase) were not significantly elevated. However, osteopontin is associated with the extracellular matrix formation and total protein is related to the strength of connection with the underneath substrate. The results indicate that the presence of the electric field influences positively cells communication with the surrounding environment, that is the neighbouring cells and the substrate. This shows that the formation of an appropriate tissue *in intro* will take place only in the presence of bioelectricity. The growth rate of the tissue will be physiologic, that is optimum, only when the laboratory micro-environmental conditions will closely biomimic the natural ones.

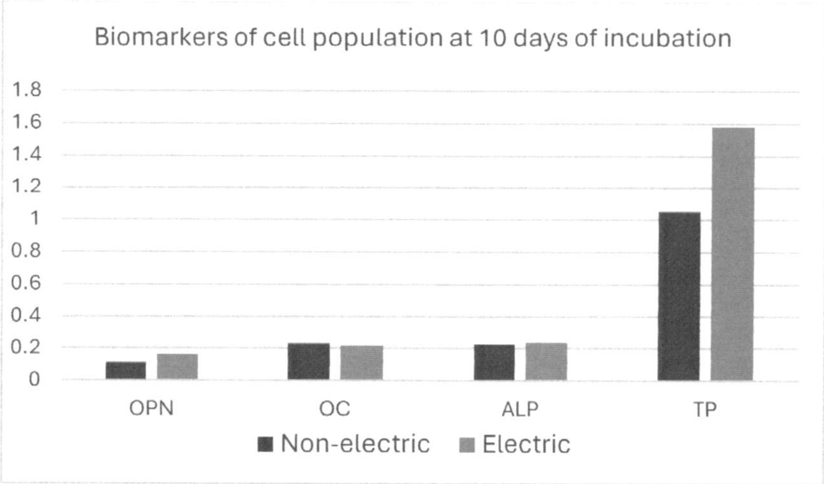

Figure 2. The values of different biomarkers (OPN, osteopontin; OC, osteocalcin; ALP, alkaline phosphatase and TP, total protein) expressed in stem cells after 10 days of incubation on 3D electrically conductive scaffolds, non-exposed (non-electric) and exposed (electric) to an electric field

2.2 Human related clinical findings

Physiologic bioelectricity is compulsory for our bodies' health. Nowadays we detain technologies to interpret, interrupt, or redirect the biologic communication signals thanks to improved methods for measuring our innate electricity at the cellular level and a resulting deeper understanding of the cellular processes. A misfunction of our bioelectricity will determine side effects such at both physical and psychologic level. For people with intense physical activity, the consumption of synthetic electrolytes is required to maintain normal bioelectricity levels and perform accordingly. On the other hand, the many types of electromagnetic waves emitted from the external environment will interfere with our physiologic bioelectricity.

The use of mobile and other electromagnetic devices affects the human body. Each body will have specific thermosensitivity; the adjustment of the body temperature to adapt to environmental conditions is managed by the cardiovascular system through the blood circulation. In the case of a high filed frequency, the system is overload and thereof it becomes impossible to maintain a steady internal temperature. Some aftereffects of intense use of mobile phones and similar devices are the biological and the psychological unbalance [11]. According to Kivrak et al. [12], electromagnetic fields can have a variety of chemical effects, such as degrading large molecules within cells and changing the ions balance. Reactive oxygen species that are produced can harm DNA, lipids, and proteins in cells. The purpose of antioxidant defense systems is to inhibit the production of free radicals and shield the biological system from their damaging effects. However, research has shown that prevention is not always feasible and that exposure to electromagnetic fields causes oxidative stress in a variety of bodily tissues.

Human awareness and perception of radiofrequency radiation have been examined in a study by Owolabi et al. [13]. Human self-reported behaviour and sleep patterns were profiled using validated and structured questionnaires. According to the study, respondents' mental health had been significantly and negatively impacted by radio frequencies. There is an urgent need to educate the public about radio frequencies and exposures, beyond the level of common knowledge and awareness that exists today.

Electromagnetic hypersensitivity (EHS) is a phenomenon that manifests as symptoms after exposure to electromagnetic fields. It is regarded as a syndrome that includes a wide range of non-specific multiple organ symptoms, including acute and chronic inflammatory processes primarily in the respiratory, cardiovascular, and musculoskeletal systems, as well as the skin and nervous systems [14]. More research provided conclusive evidence about the effects of electromagnetic fields on people. As was previously stated, there is no turning back to a less technologically advanced society, so solutions are required.

As recommended [15], the key to diagnosing and treating electromagnetic fields (EMF) side effects is to thoroughly investigate the medical history that includes all symptoms and their occurrences in spatial and temporal terms as well as in relation to electromagnetic field exposures. The prevention or reduction of electromagnetic field (EMF) exposure—that is, the elimination or reduction of all sources of high EMF exposure at home and at work—should be the main focus of treatment. Understanding the radiation protection principles of time, distance, and shielding is one of the best ways to be at low risk or no risk at all, according to the US Environmental Protection Agency (EPA) [16]. In terms of scientific progress related to the subject, more *in vitro* to clinical investigation studies are required to understan in depth the extent of affections caused by exposure on tissue, organs and organism and to elaborate personalized solutions.

3. Conclusions

The present manuscripts discussed the impact of an electric or electromagnetic field on the body, starting from human cells to whole organism. The presence of an appropriate electric field in the body is compulsory; the lack of balanced metabolism leads to the need of applying external solutions such as the administration of electrolytes.

Bioelectricity of certain ranges is essential for the human health; this have been proved in laboratory investigations with human cell cultures. Several types of cell populations such as the stem and the bone cells showed improved *in vitro* functionality in the presence of an electric field. The generation of an appropriate electric field is the key parameter for enabling a normal tissue growth rate in lab conditions. Consequently, the functionalization of biomaterials for implants and artificial organ components should also be enabled through conferring them electrical properties that allow charge transfer from the biologic to the synthetic tissue and vice-versa.

External electromagnetic fields that exist in the surrounding environment interfere with the inner electricity. The direct effect is the heating of the tissues. In case of long-term exposure this may result in serious pathologies. However, there are ways to reduce the radiation around us, to predict, prevent and treat the effect of radiation. As reduction of technologies is impossible, personalized solutions based on thorough *in vitro* and *in vivo* investigations are demanded. Finally, educating citizens towards this direction is extremely important to establish a common vision that will be in the benefit of technological advancement and at the same time, will provide sufficient protection from medical viewpoint.

Acknowledgements

This present investigation was performed under grant agreement No 101112347, project NerveRepack (Intelligent neural system for bidirectional connection with exoprostheses and exoskeletons) and supported by the CHIPS Joint Undertaking and its members.

References

[1] Kwang Suk Park, Humans and Electricity: Understanding Body Electricity and Applications, Springer Nature, Apr 3, 2023, Technology & Engineering, 408 pages
[2] https://www.who.int/news-room/questions-and-answers/item/radiation-electromagnetic-fields#:~:text=The%20main%20effect%20of%20radiofrequency,can%20be%20harmful%20to%20health.
[3] Kawthar A. Diab, The Impact of the Low Frequency of the Electromagnetic Field on Human, Adv Exp Med Biol - Cell Biology and Translational Medicine (2019) 7: 135–149
[4] Anamta Sohail, Ahsan Ali, Hamna Shaukat, Farah Mukhtar Bhatti, Shaukat Ali, Sallam A. Kouritem, Mohammad Noori, Wael A. Altabey, Integrating self-powered medical devices with advanced energy harvesting: A review, Energy Strategy Reviews, Volume 52, 2024, 101328
[5] Ionescu, O.N.; Franti, E.; Carbunaru, V.; Moldovan, C.; Dinulescu, S.; Ion, M.; Dragomir, D.C.; Mihailescu, C.M.; Lascar, I.; Oproiu, A.M.; et al. System of Implantable Electrodes for Neural Signal Acquisition and Stimulation for Wirelessly Connected Forearm Prosthesis. Biosensors 2024, 14, 31. https://doi.org/10.3390/bios14010031
[6] Meng S, Rouabhia M and Zhang Z (2011) Electrical Stimulation in Tissue Regeneration. Applied Biomedical Engineering. InTech. Available at: http://dx.doi.org/10.5772/18874.
[7] Richard Nuccitelli, Pamela Nuccitelli, Samdeo Ramlatchan, Richard Sanger, Peter J.S. Smith, Imaging the electric field associated with mouse and human skin wounds, Volume16, Issue 3, May–June 2008, Pages 432-441.
[8] Diana V. Portan, Despina D. Deligianni, George C. Papanicolaou, Vassilis Kostopoulos, Georgios C. Psarras, Minos Tyllianakis, "Combined Optimized Effect of a Highly Self-Organized Nanosubstrate and an Electric Field on Osteoblast Bone Cells Activity", BioMed Research International, vol. 2019, Article ID 7574635, 8 pages, 2019. https://doi.org/10.1155/2019/7574635
[9] Chatzistamatiou, T.K.; et al., Optimizing isolation culture and freezing methods to preserve Wharton's jelly's mesenchymal stem cell (MSC) properties: an MSC banking protocol validation for the Hellenic Cord Blood Bank, Transfusion 2014, 54(12), 3108-3120.
[10] Kozaniti, F.K.; Manara, A.E.; Kostopoulos, V.; Mallis, P.; Michalopoulos, E.; Polyzos, D.; Deligianni, D.D.; Portan, D.V. Computational and Experimental Investigation of the Combined Effect of Various 3D Scaffolds and Bioreactor Stimulation on Human Cells' Feedback, Appl. Biosci. 2023, 2, 249-277.
[11] L. Vijayalakshmi, P. Nirmala Devi, Impacts of RF radiation from mobile phones on human health and its remedies, J. appl. res. technol vol.18 no.5, 269-278.
[12] Kıvrak EG, Yurt KK, Kaplan AA, Alkan I, Altun G. Effects of electromagnetic fields exposure on the antioxidant defense system. J Microsc Ultrastruct. 2017 Oct-Dec;5(4):167-176.
[13] Owolabi J, Ilesanmi OS, Luximon-Ramma A. Perceptions and Experiences About Device-Emitted Radiofrequency Radiation and Its Effects on Selected Brain Health Parameters in Southwest Nigeria. Cureus. 2021 Sep 23;13(9):e18211
[14] Kaszuba-Zwoińska J, Gremba J, Gałdzińska-Calik B, Wójcik-Piotrowicz K, Thor PJ. Electromagnetic field induced biological effects in humans.
[15] Belyaev I, Dean A, Eger H, Hubmann G, Jandrisovits R, Kern M, Kundi M, Moshammer H, Lercher P, Müller K, Oberfeld G, Ohnsorge P, Pelzmann P, Scheingraber C, Thill R. EUROPAEM EMF Guideline 2016 for the prevention, diagnosis and treatment of EMF-related health problems and illnesses. Rev Environ Health. 2016 Sep 1;31(3):363-97
[16] https://www.epa.gov/radiation/protecting-yourself-radiation

Study upon the influence of mobile phone protective case on SAR distribution

Petronela-Camelia Oprea[1]*, Marius Păuleț[1] and Alexandru Sălceanu[1]

[1] Faculty of Electrical Engineering, "Gheorghe Asachi" Tehnical University of Iași, Romania

*E-mail: petronela-camelia.oprea@student.tuiasi.ro

Abstract. The article presents a series of simulations performed with CST Studio software, using the Hugo male model, a member of CST Voxel Family. These simulations aim to study the effect that the conductivity (thermal and electrical) of the phone protective case might have on the SAR distribution. In the first part, details are presented on how to understand and use the SAR parameter, on how CST Studio simulates it, complementary with the method of delivery of the obtained data. All 6 communication ports of the available phone model have been tested, 4 of them intended for 2G, 2G+, 3G, 4G mobile phone technologies, the other 2 being dedicated to the two Wi-Fi standards (2.4 and 5 GHz). The phone has been placed in the left breast pocket (near heart). The SAR values have been compared for three different cases: silicone protective case without conductive foil, with 0.5 mm Aluminum conductive foil, placed between the phone and the case and with foil of the same thickness, but made of Copper. Conclusions have been drawn regarding the SAR reduction percentages as a result of the use of conductive foils. A recommendation regarding an efficient and accessible solution for designing and making phone protective cases has been issued.

1. Introduction

Our daily life takes place in smart homes, offices, cities or, all in all, in smart environments. The design and management of these modern built environments from all points of view (economic, energetic, safety and health) occupies a priority place in the current research agenda of the European Union. In this multitude of related issues, influencing and conditioning each other, a special place belongs to the cumulative study of the effects that the electromagnetic smog associated with contemporary urban agglomerations could have on human health. Applications that generate and/or use non-ionizing electromagnetic radiation are extremely numerous. It would be virtually impossible to imagine our daily lives without wireless communications. The only possible response of the most diverse collectives of researchers around the world is the continuation and development of studies on measuring and monitoring the level of exposure to the multitude of electromagnetic emissions from the perspective of possible impact on human health. Any scientific approach is based on the definition and acceptance of some metrics and indicators that should allow comparisons establishing red lines, those thresholds whose crossing must worry and alarm us. One of the widely accepted indicators used to assess the level of exposure to non-ionizing radiations (mainly associated with mobile cell telephony) is the Specific Absorption Rate (SAR).

SIMULIA CST Studio Suite [1] provides a comprehensive set of tools that enables device manufacturers to verify the compliance with the regulations dedicated to consumers' health and safety. There are an important number of prohibitions and regulations that limit the direct measurements of electromagnetic fields inside the organs and tissues of a living human organism. In vivo tests are strictly limited, in vitro tests have very low relevance, the highest expectations are related to "in silico" tests. More precisely, it starts from 2D images, cross-sectional "slices" of the human body, obtained by Magnetic Resonance Imaging (MRI) or Computer Tomography (CT). By extending from pixels associated with 2D images to 3D volumetric elements called voxels, the so-called "voxel-based human models" have been obtained. One of these models is "male Hugo", developed in the CST Voxel Family, using the achievements of the "Visible Human Project®" In this way, the heating produced by the impact of electromagnetic waves on the body can be simulated, i.e. the SAR distribution.

Another widely used model is VIP-Man, developed by the National Research Center for Environment and Health in Germany [2-3].

Interesting studies have been focused on the evaluation of SAR when the patient is placed in an MRI machine. Relevant comparisons have been performed between different human models: male, female, child or even newborn [4]. A useful study has been performed on the SAR distribution in the body of a child in a classroom with many laptops [5]. Authors Zuzana Psenakova et. all [6] simulated the SAR values on a fetus in the womb.

Among the main factors influencing the correctness of the SAR distribution, the accuracy of the human model plays an extremely important role. The Hugo model used in our study has a precision of 5x5x5 mm^3, which is high enough for conclusive results.

The CST simulation environment works according to the IEEE/IEC62704-1 Standard [7].

2. Methodology

In our study, we used the CST Studio Suite simulation environment in the full-wave time domain as in Figure 1. The excitation pulse has a duration of 1.184849 ns.

The specific absorption rate (SAR) is defined as the energy absorbed by the human body related to a certain mass and time, according to equation 1:

$$SAR = \frac{d}{dt}\left(\frac{dW}{dm}\right) = \frac{d}{dt}\left(\frac{dW}{\rho dV}\right) \quad (1)$$

Equation 2 shows a different approach leading to the same result, that is possible when the intensity of the electric field induced in the body is known:

$$SAR = \frac{\sigma}{\rho}E^2 \quad (2)$$

where: σ – tissue conductivity [S/m], ρ – tissue mass density [kg/m^3], E – RMS electric field intensity [V/m].

Essentially, SAR represents the measure of the amount of NIR energy transformed into thermal energy that is absorbed by the human body during the use of a specific device.

If in other respects there are differences in approach between the European guidelines (ICNIRP) and the American ones (IEEE and Federal Communications Commission), in terms of the maximum acceptable limits for SAR there is consensus [8]: 2 W/kg or 2 mW/g in any 10 g of tissue (in our case, skin and tissue around the heart area, near the breast pocket where the phone is usually placed). Without complying with this condition, the phone cannot be sold on the market.

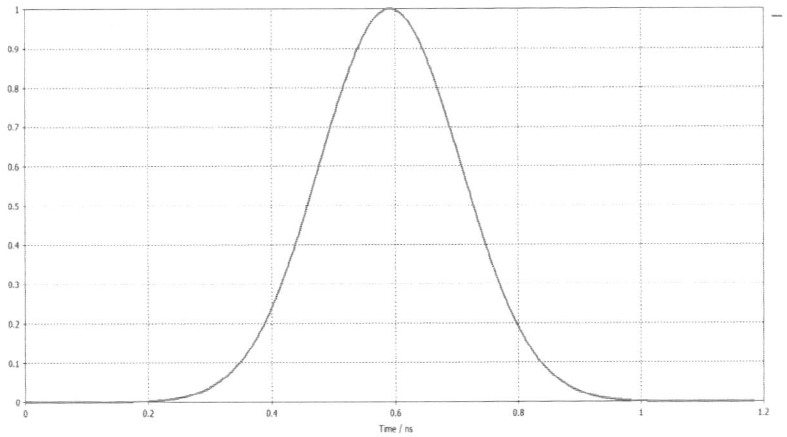

Figure 1. Excitation signal

3. Modeling

3.1. Introducing the CST Studio Suite simulation environment

CST Studio Suite is a set of advanced software tools used for simulation and analysis in various applications [10]. In Figure 2 is presented a flowchart showing the steps to be followed in order to obtain a good simulation with the help of CST Studio software.

It is also important to note that SAR simulation in the CST Studio Suite program involves special attention to the geometric details of the model and the selected parameters.

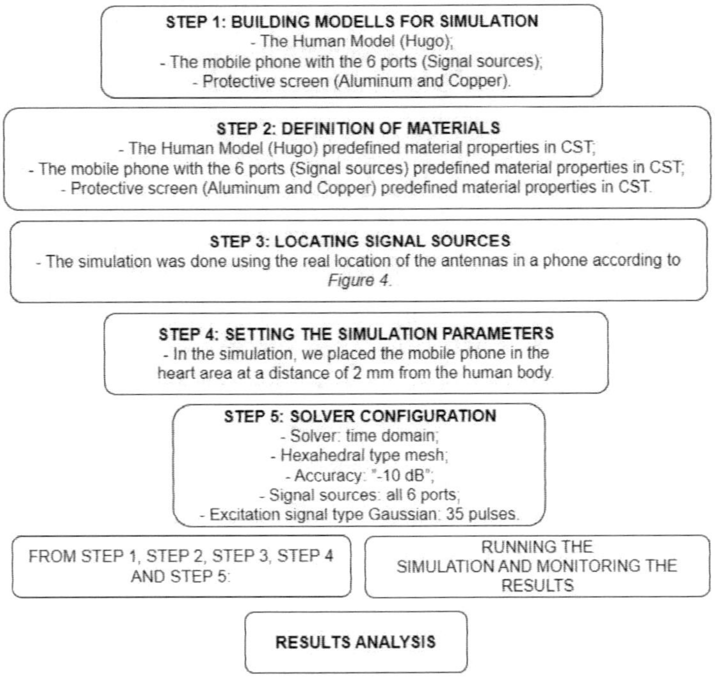

Figure 2. The necessary steps to create a good simulation in CST Studio

IOP Conf. Series: Materials Science and Engineering **1320** (2024) 012022 doi:10.1088/1757-899X/1320/1/012022

3.2. Model development in CST Studio Suite

For the human body we used a predefined model in the CST simulation environment, namely the Hugo model: gender-male, height-180 cm, weight-103 Kg, age-38 years, as shown in Figure 3.

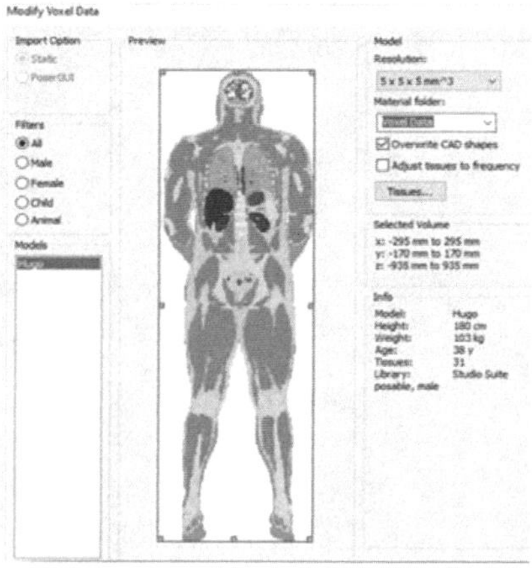

Figure 3. Male Hugo model in CST Studio Suite

The electrical, magnetic, thermal and mechanical properties of all the organs and tissue have been defined. Of primary interest for the SAR distribution study are the skin properties, at voxel level:

- Type: Normal
- Epsilon: 69.45
- Electric cond.: 0.507 [S/m]
- Magnetic permeability: 1
- Rho: 1100 [kg/m^3]
- Thermal cond.: 0.3 [W/K*m]
- Specific heat: 3500 [J/K*kg]
- Diffusivity: $7.79221*10^{-8}$ [m^2/s]
- Bloodflow: 9000 [W*K*m^3]

The signal sources used in the simulation are the 6 ports of the mobile phone. It should be considered the cumulative effect of the quasi-simultaneous possible operation of 2 or 3 ports, resulting in higher SAR values.

The destination for the 6 ports is here briefly presented:

Port 1 is for E(extended)-GSM (Global System for Mobile telecommunications). It works around the frequency of 900 MHz. More precisely, it means an addition of 10 MHz at the lower side of the primary GSM 900 MHz band. In conclusion, E-GSM ranges from 880 - 890 MHz and from 925 - 935 MHz.

It uses FDD transmission/reception mode (with split frequencies, 2 frequencies are used, one for transmission, one for reception, so it can work simultaneously, increasing data transmission speed).

Port 2 and port 3 are for DCS (Digital Cellular System) and GSM-1800 type, the carrier frequency being for both 1800 MHz, also using FDD transmission/reception mode.

Port 4 and port 5 are Wi-Fi ports that work at the frequency of 2,4 GHz and 5 GHz respectively.

Port 6 is an E-UTRA 7 (Evolved Universal Terrestrial Radio Access) port. E-UTRA is the air interface that made possible the upgrade from 3G (UMTS) to 4G (LTE) technology. There are 108 assigned operating bands. Among these, Band 7 uses frequency-division duplexing (FDD) mode, while the uplink and downlink frequencies are around 2600 MHz.

The approximate location of the 6 communication ports is shown in Figure 4:

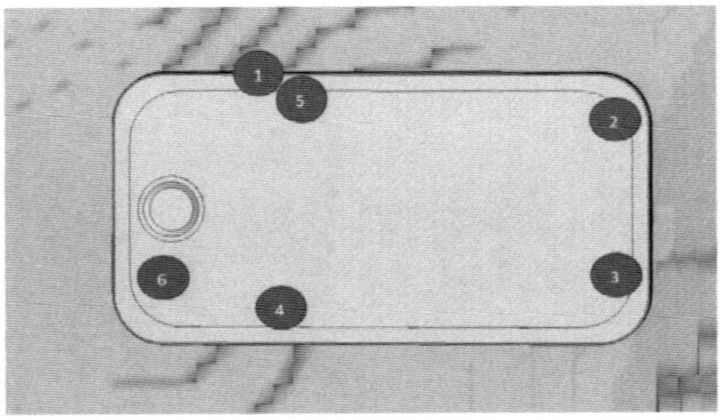

Figure 4. Location of the 6 communication ports on the mobile phone

The program uses for simulation the electrical diagram presented in Figure 5. The cut-off frequency of the high-pass filters is correlated with the carrier frequency used by the technology associated with the respective port.

Port 1 (E-GSM) is simulated using an LC High pass filter, L=2.5 nH and C=4 pF.

Ports 2 and 3 (DCS-GSM) are simulated by an RC, also High pass filter where the values of the two capacitors have different values: for port 2, C=3.3 pF and for port 3, C=1.2 pF.

Port 4 (Wi-Fi 1) has no filter.

Port 5 (Wi-Fi 2) is simulated by an LC High pass filter, C=12 pF and L=3.6 nH.

Port 6 (E-UTRA 7) is also simulated by a LC High pass type filter, C=2.2 pF and L=3.9 Nh

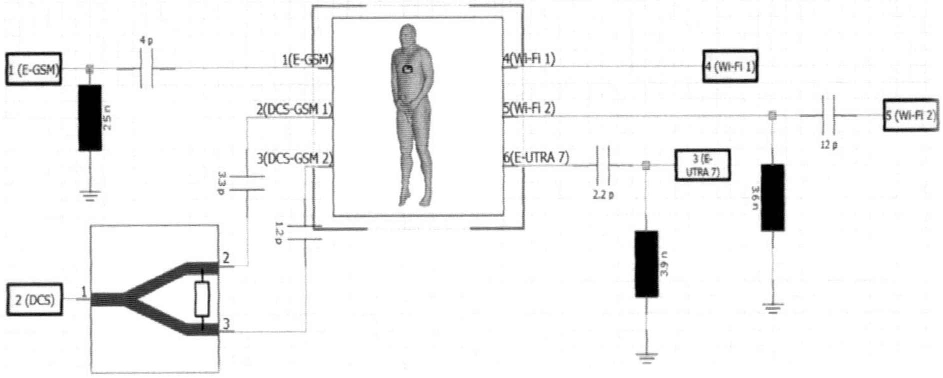

Figure 5. Hugo model simulation wiring diagram implemented in CST Studio Suite

3.3. Simulating the influence of a conductive screen (foil) on SAR distribution

Most smart mobile handlers use a drop protecting case, shock absorbing and scratch resistant. This back case is usually made of silicone, a non-conductive material. In this paper, we present the results of our simulations when between the rear of the phone and the protective case we additionally inserted a copper or aluminum foil, with a thickness of 0.5 mm. We compared the results obtained for SAR distribution around the proximity of the heart, when the phone is kept in the chest pocket of the coat, (Figure 6) for the following three distinct situations: without additional conductive foil, with copper foil, respectively with aluminum foil. We chose to study the influence of these metallic screens on SAR distribution because they are good conductive materials, both electrical and thermal.

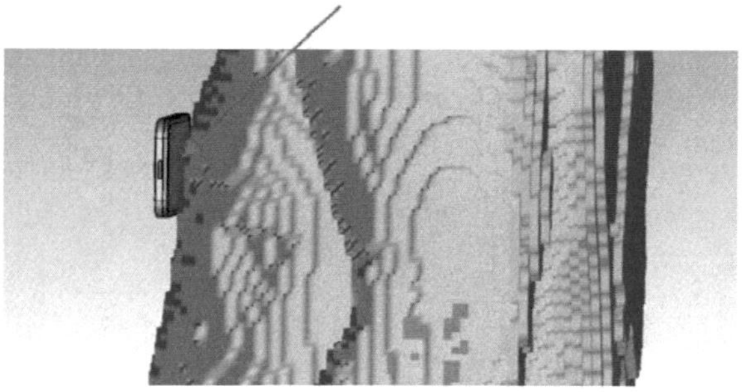

Figure 6. Positioning the Aluminum/Copper protective screen

For Aluminum the following properties have been defined in CST:

- Type: Lossy metal
- Magnetic permeability: 1
- Electric cond.: $3.56*10^7$ [S/m]
- Rho: 2700 [kg/m^3]
- Thermal cond.: 237 [W/K*m]
- Specific heat: 900 [J/K*kg]
- Diffusivity: $9.75309*10^{-5}$ [m^2 /s]
- Young's modulus: 69 [kN/mm^2]
- Thermal expansion: 23 [10^{-6} /K]

For Copper the following characteristics have been defined:

- Material Set: Default
- Type: Lossy metal
- Magnetic permeability: 1
- Electric cond.: $5.96*10^7$ [S/m]
- Rho: 8930 [kg/m^3]
- Thermal cond.: 401 [W/K*m]
- Specific heat: 390 [J/K*kg]
- Diffusivity: 0.000115141 [m^2 /s]
- Young's modulus: 120 [kN/mm^2]
- Poisson's ratio: 0.33
- Thermal expansion: 17 [10^{-6} /K]

IOP Conf. Series: Materials Science and Engineering **1320** (2024) 012022 doi:10.1088/1757-899X/1320/1/012022

4. Simulation results and their interpretation

We have successively simulated the operation (communication) at each of the six ports. Naturally, we have chosen for every port the frequency of the carrier mainly used by the respective technology. The distance between the phone and the Hugo's skin has been considered 2 mm.

After the simulation performed, CST provides images in which the color code suggests very intuitively the temperature distribution, an example being offered by Figure 7.

In the same time, it also provides numerical data, which allow extracting the maximum and minimum values for SAR.

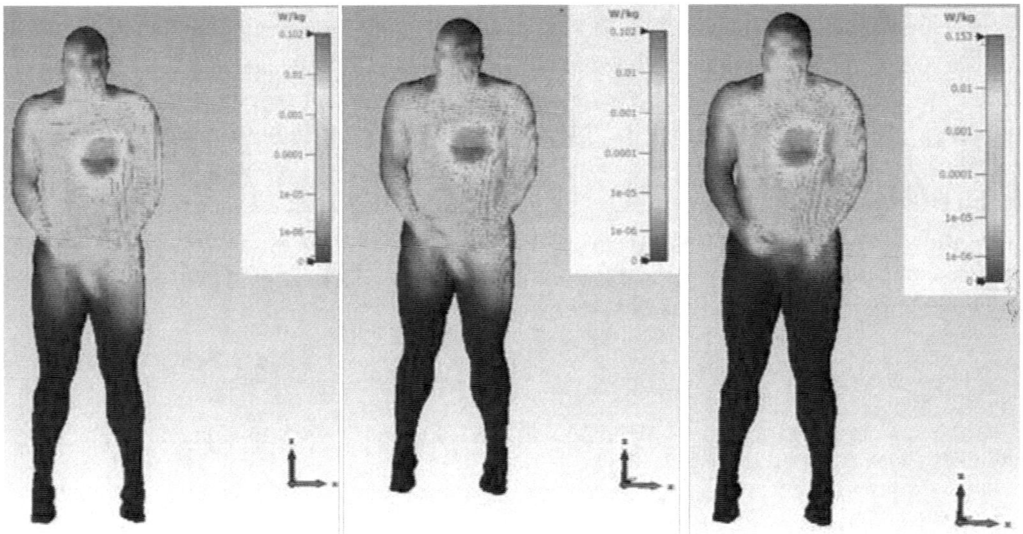

Figure 7. Example of SAR distribution around the heart zone for Aluminium screen, Copper screen and without screen by P [1] E-GSM 900 MHz

To each communication port of the phone with the corresponding carrier frequency, we have associated a row in Table 1.

The maximum values recorded for SAR (the three compared situations) have been introduced.

Table 1. Values of the max SAR parameter

Ports	Frequency [MHz]	SAR max [W/kg]			SAR percentage decrease [%]
		Without screen	Al Screen	Cu Screen	
P [1] E-GSM-900	900	0.152856	0.102143	0.102144	66.823677
P [2] DCS	1800	0.162305	0.119182	0.119189	73.430881
P [3] GSM-1800	1800	0.148092	0.113614	0.113623	76.719875
P [4] Wi-Fi 1	2400	0.553150	0.503079	0.503288	91.301252
P [5] Wi-Fi 2	5000	0.476275	0.239876	0.239891	50.368169
P [6] E-UTRA 7	2600	0.256880	0.220714	0.220822	85.921052

IOP Conf. Series: Materials Science and Engineering **1320** (2024) 012022 doi:10.1088/1757-899X/1320/1/012022

5. Conclusions

The simulation environment CST Studio together with the human model Hugo, gives us a very close to reality evaluation of the phenomena that can occur during the use of a mobile phone. Due to the very precise information that computer tomography and the magnetic resonance imaging provide about the organs and tissues of the human body, the results of the simulations present a very high degree of accuracy.

All maximum values recorded for SAR are considerably lower than the accepted limit of 2 W/kg. This does not mean, however, that possible SAR reduction measures are not justified. It has been proven that the use of a conductive foil placed between the rear face of the phone and its protective case leads to a decrease in SAR to 50-91% of the value without protection. The simulations have also showed completely insignificant differences between the case of copper and aluminum foil. A conclusion we expected, the good thermal and electric conductive properties of the two metals being very close. Anyway, a possible solution to reduce SAR could be the production and sale of silicon casings, but with good conductive properties obtained as a result of appropriate copper or graphite wire insertion.

References

[1] https://space.mit.edu/RADIO/Documentation/CST%20Studio%20Suite%20-%20Getting%20Started.pdf/ Accessed on 27 March 2024, CST Studio Suite Guide;
[2] Min ZHANG et all 2010, Asia Pacific International Symposium on Electromagnetic Compatibility, Beijing, China, "Influence on SAR due to Metallic Frame of Glasses based on High Resolution Chinese Electromagnetic Human Model";
[3] XG Xu et all 2000, Health Physics: The Radiation Safety Journal 78(5): p 476-486. "VIP-MAN: an image based whole-body adult male model constructed from color photographs of the Visible Human Project for multi-particle Monte Carlo calculations";
[4] AS Emhemmed et all 2023, IEEE 3rd International Maghreb Meeting of the Conference on Sciences and Techniques of Automatic Control and Computer Engineering (MI-STA), Benghazi, Libya, pp. 750-754, "Evaluation and Simulation of SAR Among Patients Using MRI Machines";
[5] NE Soares et all 2023, IEEE MTT-S Latin America Microwave Conference (LAMC), San José, Costa Rica, pp. 58-60,"SAR Estimations in a Child Due to RF Exposures from Several Laptops in a Classroom Environment";
[6] Z. Psenakova et all 2020, ELEKTRO, Taormina, Italy, pp. 1-6, "Investigation of Specific absorption rate (SAR) near model of fetus in uterus";
[7] IEC/IEEE International Standard for Determining the Peak Spatial Average Specific Absorption Rate (SAR) in the Human Body from Wireless Communications Devices, 30 MHz – 6 GHz. Part 1: General Requirements for using the Finite Difference Time Domain (FDTD) Method for SAR Calculations;
[8] ICNIRP Guidelines for limiting exposure to time-varying electric, magnetic and electromagnetic fields (up to 300 GHz), published in: Health Physics 74 (4):494-522, 1998;
[9] https://www.fcc.gov/ Accessed on 27 March 2024, Federal Communication Commission, Specific Absorption Rate (SAR) For Cell Phones What It Means for You, Washington;
[10] https://www.3ds.com/products/simulia/cst-studio-suite/Accessed on 27 March 2024.

IOP Conf. Series: Materials Science and Engineering **1320** (2024) 012023 doi:10.1088/1757-899X/1320/1/012023

Spectrogram based Wi-Fi usage activity classification using deep learning

Annamaria Sarbu[1]*, Simion Şuhani[1], Mirela Şorecău[1], Emil Şorecău[1]

[1]Sibiu, "Nicolae Balcescu" Land Forces Academy
*paljanosanna@yahoo.com

Abstract The article presents and characterizes the ability of a neural network to distinguish between different Wi-Fi activities based on their distinct spectral features. To achieve this objective, we have built a database containing spectrogram images retrieved by over-the-air measurements of signals emitted in 802.11ax Wi-Fi networks deployed in both the 2.4 GHz and 5 GHz frequency bands. The dataset consists of spectrograms labeled by six distinct user activities, namely file download, speed test, video streaming, file upload, video call, and voice over IP call (VoIP). The network training parameters such as learning rate and validation frequency were optimized to enhance model performance. The influence of different activation functions, ReLU, leaky ReLU, eluLayer, and swishLayer from Matlab was also evaluated to achieve neural network fine-tunning. Results demonstrate the effectiveness of the proposed neural network architecture in accurately classifying Wi-Fi usage patterns, achieving test accuracies above 98%.

1. Introduction

The demand for seamless wireless connectivity in today's digital landscape has made wireless fidelity (Wi-Fi) networks omnipresent [1]. Nowadays, Wi-Fi plays an important role in various aspects of modern life, such as communication [2], entertainment [3], healthcare [4] and education [5].

In this context, understanding and analyzing Wi-Fi usage activity patterns can provide valuable information for optimizing network performance, ensuring network security, and managing bandwidth effectively. The authors of [6] propose the use of the Wi-Fi network association information as a basis for the design of intelligent systems for smart buildings. They evaluated the capacity of machine learning (ML) models to accurately predict Wi-Fi access point demand for energy-efficient smart buildings. This idea could be further developed to include security related aspects. For example, extensive bandwidth consuming activity in a specific area (access point range) could indicate a flooding attack aimed to congest the network and degrade user quality of service.

From another point of view, we appreciate that the classification of Wi-Fi usage activities could enhance the efficiency of deploying Unmanned Aerial Vehicles (UAVs) as floating access points (APs)[7]. By accurately identifying Wi-Fi usage patterns, such as peak usage times, specific applications being utilized, and areas experiencing network congestion, UAVs equipped with this knowledge can dynamically adjust their movements to cover hot zones and alleviate network congestion more effectively. For example, if the classification model identifies a particular area experiencing high video streaming activity during certain times of the day, UAVs can be directed

IOP Conf. Series: Materials Science and Engineering **1320** (2024) 012023 doi:10.1088/1757-899X/1320/1/012023

to hover over that area during those times to provide additional bandwidth and improve user experience. This targeted deployment based on Wi-Fi user service type can optimize the utilization of resources and enhance the overall performance of the network.

Recently, the authors of [8] proposed the use of Wi-Fi usage data along with mobile device data and Bluetooth as behavioral biometric traits. To this extent, integrating Wi-Fi usage activity classification could provide valuable insights into user behavior and habits, enhancing the accuracy and robustness of behavioral biometric systems.

Deep learning techniques, particularly neural networks, offer promising avenues for automatic and precise classification in various fields [10-14] including different Wi-Fi signal types/characteristics. The authors of [11] propose the use of spectrogram images in conjunction with a convolutional neural network to classify a malware file to its corresponding family and to differentiate it from a benign file. Spectrogram type representations are used in conjunction with neural networks to classify audio signals [12], ECG signals [13] and even mobile application emission type [14].

The main objective of this article is to develop and train a neural network to classify spectrogram images according to their corresponding Wi-Fi usage activity. To our knowledge, this study represents the first attempt to apply a deep learning algorithm for Wi-Fi usage activity classification, relying solely on data obtained from spectrum measurements. Spectrogram images were recorded and labeled according to six types of user activities by conducting over the air measurements in a realistic electromagnetic environment. The spectrogram image dataset is used to train the designed network and its performance is evaluated with respect to the training parameters and the activation layer functions.

2. Materials and methods

2.1 Data collection and pre-processing

The training, validation and test data were collected in the form of spectrogram images of signals emitted in two 802.11ax Wi-Fi networks operated in the 2.4 GHz and 5GHz frequency band, respectively. Wi-Fi communication was established between a user device (laptop) and the Tp-Link Archer AX53 router model. The router was configured to use 40 MHz wide channels (maximum real-time bandwidth of the SA), specifically channel 2 for the 2.4GHz band (center frequency of 2.437 GHz) and channel 46 for the 5 GHz band (center frequency of 5.23 GHz). The measurements were performed indoors with other wi-fi networks being available and in use at the time of measurements, reflecting typical usage conditions in real-world operating scenarios. Six types of actions were initiated from up to three user devices, namely: file download, speed test, video streaming, file upload, video call and voice over IP call (VoIP). Over 6500 spectrograms were retrieved in total and labeled according to the network type and usage scenario.

Over the air measurements were conducted by means of a system comprised of an OmniLOG 70600 antenna, a real-time Spectrum Analyzer (model R&S®FSVR) and a computer used for process automatization. The remote control of the SA and retrieval of spectrograms were performed through a custom developed Python script. The collected spectrograms were stored as RGB images, each approximately 250 kB in size, onto a local hard drive with a .jpg extension. The

IOP Conf. Series: Materials Science and Engineering **1320** (2024) 012023 doi:10.1088/1757-899X/1320/1/012023

experimental setup for data acquisition is illustrated in figure 1. A similar experimental setup was used in [15] where it is also detailed.

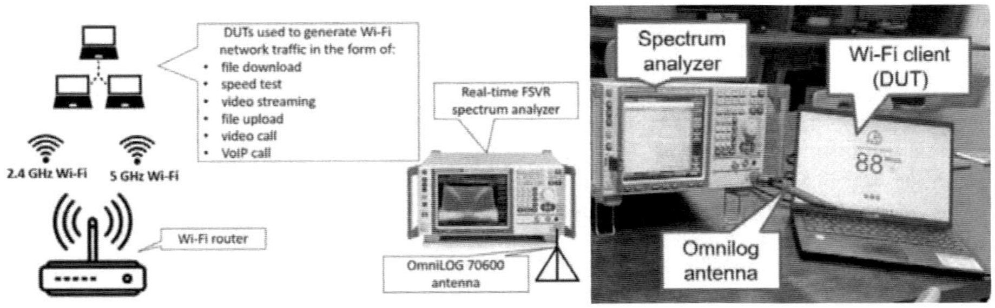

Figure 1. Experimental setup

The RGB images, consisting of three channels, were cropped to a size of 955x1584, focusing solely on the colored region of the image, and then imported into a MATLAB datastore. The images were cropped to decrease their size for training the network, thereby optimizing the training process by shortening its duration. Figure 2 presents a spectrogram corresponding to file download in the 5 GHz Wi-Fi (a) and a cropped spectrogram corresponding to VoIP in the 2.4 GHz Wi-Fi network (b).

We can observe that there is a difference between the channel occupancy for the two use cases. For file download the entire 40 MHz Wi-Fi channel appears to be heavily utilized as compared to the VoIP case when the spectrogram reveals a different usage pattern. This difference stems from the technical characteristics of each application. For file downloads, the transmission typically involves continuous and high-volume data transfer, resulting in sustained channel usage. On the other hand, VoIP communication operates with sporadic bursts of data, leading to intermittent channel utilization and distinct spectral signatures. Video streaming involves a continuous flow of data, utilizing substantial bandwidth over longer durations, much like file download. Video calls, resembling VoIP, consist of sporadic data bursts, occupying the entire bandwidth and featuring more uniform usage patterns than VoIP. Speed tests is associated with brief but intense bursts of data transmission, leading to transient yet significant channel occupancy.

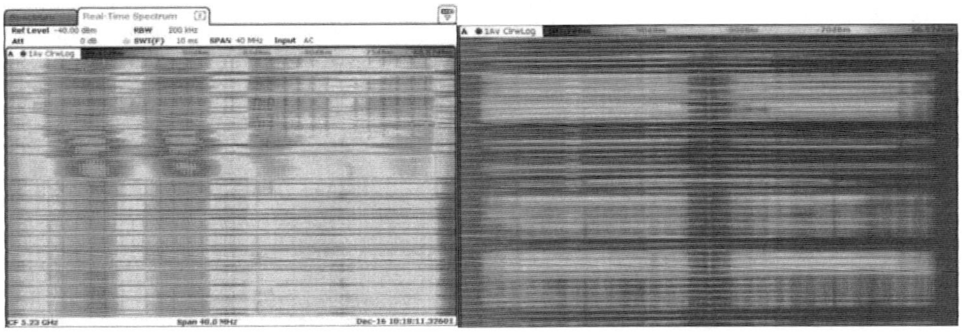

(a) (b)

Figure 2. File download (a) and VoIP (b) spectrogram retrieved by the SA

The images within the datastore were divided into 70% for training, 20% for validation, and 10% for testing. Table 1 provides an overview of the total number of spectrogram images acquired for each usage scenario while figure 3 displays a histogram illustrating the distribution of training images within both 2.4 GHz and 5 GHz datastores.

Table 1. Total number of spectrogram images

	Wi-Fi network	2.4 GHz	5 GHz
		Count	Count
	file download	500	588
	speed test	500	493
Label	video streaming	600	603
	file upload	550	637
	video call	550	539
	VoIP	500	490

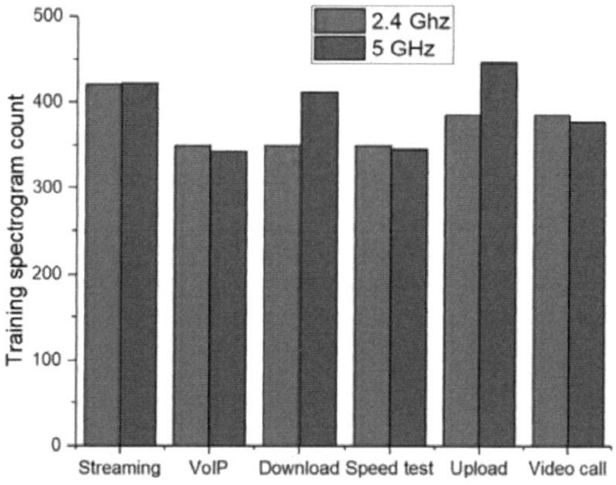

Figure 3 Training dataset histogram

2.2 Neural network design

In this study, we developed and trained a supervised neural network in Matlab. The neural network architecture is presented in figure 4.

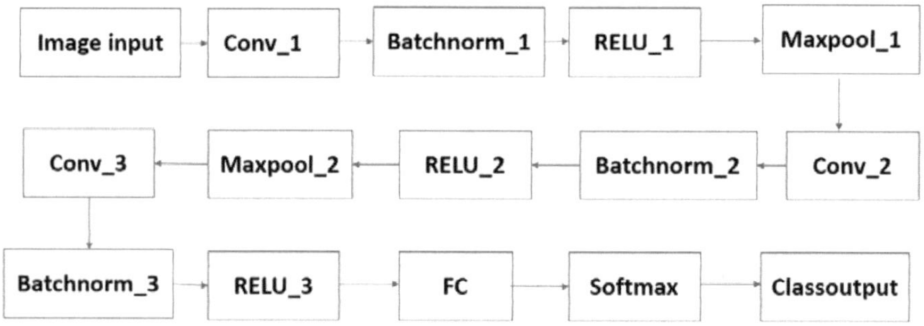

Figure 4 Deep learning network architecture

The image input layer serves as the initial entry point, accepting the RGB 955x1584x3 sized images from the constructed datastore. Convolutional layers extract features from the input images through convolution operations, while batch normalization layers normalize the outputs for stable training. ReLU layers utilize the rectified linear unit activation function to introduce non-linearity, enhancing the model's ability to capture intricate patterns. Maximum pooling layers extract the dominant features and reduce the number of feature maps to decrease computational complexity. The fully connected layer integrates high-level features for classification, leading to a compact representation of the input data. Finally, the SoftMax layer computes the probabilities of different classes, facilitating the classification decision, while the classification output layer provides the final classification output.

The adaptive moment estimation (ADAM) optimizer was selected for training the network due to its widespread use in machine learning, particularly for deep neural networks. This variant of stochastic gradient descent (SGD), a widely used optimization algorithm, dynamically adjusts learning rates for each parameter, combining the benefits of both Adaptive Gradient Algorithm (AdaGrad) and Root Mean Square Propagation (RMSProp) optimizers. ADAM keeps track of the temporal change rate of the values involved in the learning process. These calculations are used by the algorithm to adjust the learning rate for different parts of the problem. Moreover, the algorithm incorporates bias correction for improved estimate accuracy. Compared to traditional SGD and other optimizers, ADAM tends to converge faster and more reliably, especially in scenarios involving noisy gradients or sparse data. The initial training configurations are detailed in Table II, with the remaining options set to MATLAB's default values.

Table 2. Training options for ADAM solver

Training option	Value
InitialLearnRate	0.01
LearnRateSchedule	Piecewise
LearnRateDropFactor	0.2
LearnRateDropPeriod	1
MaxEpochs	4
Shuffle	Every epoch
MiniBatchSize	32
ValidationFrequency	5
Validation patience	3
Output Network	Best validation lost

To optimize the achieved accuracy, the learning rate, the mini-bath size and the validation frequency were varied from the initial setup presented in table 2. The network was trained on a 64-bit Windows 10 machine with the following hardware resources: i7-8700 CPU 3.2 GHz, 32 GB installed RAM and 16GB NVIDIA GeForce RTX 3060.

3. Results and discussion

Figure 5 presents the training progress for the network architecture presented in figure 4 with the training parameters from table 2 applied to the 2.4 GHz datastore. One can observe that the training progress has finished successfully, passing the entire training dataset (2240 images) divided in training samples of 32 images (mini-batch size) through all 280 iterations (70 iterations/epoch x 4 epochs). Given the values of the validation frequency and validation patience, the training process finished all available iterations and did not stop to prevent overfitting. The achieved training accuracy was 99.51% while the validation and test accuracies were 97.81% and 98.75% respectively. Similar results were obtained for the network trained on the 5 GHz datastore (train accuracy 99.49%, validation accuracy 97,47% and test accuracy 97,61%). The small and subtle improvement in the test accuracy compared to the validation accuracy can be attributed to the smaller test data (10%) compared to the validation data (20%).

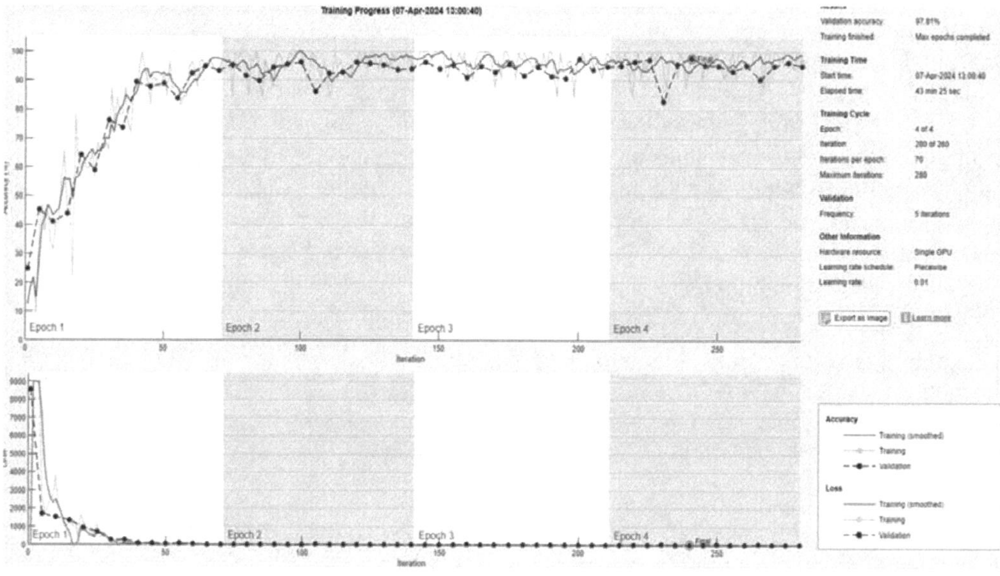

Figure 5 Network training progress

Figure 6 presents the confusion matrix for both 2.4 GHz and 5 GHz test datasets. The observed deviations from ideal network performance can be attributed to the presence of small confusions among distinct user activities, namely file download, speed test, streaming, and occasionally upload operations. This subtle confusion could stem from the overlapping characteristics of these activities, such as similar data transfer patterns or signal variations, thereby challenging the network's ability to precisely differentiate between them. Moreover, we must keep in mind that the dataset used for training and evaluation was gathered under authentic Wi-Fi usage conditions. Consequently, the spectrograms used for training and testing

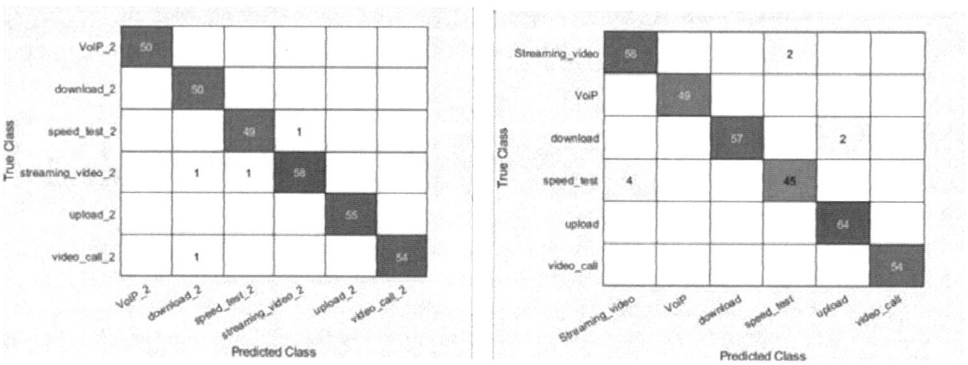

(a) (b)

Figure 6 Confusion matrix for 2.4 GHz(a) and 5 GHz(b) test datasets were labeled based on the predominant user activity, without imposing restrictions on background applications that might have been concurrently active.

The above results demonstrate the ability of the proposed network to successfully classify between Wi-Fi usage actions, validating the effectiveness of the ADAM solver in facilitating learning from spectral data. In the following subsections, we will analyze the influence of training parameters and modified ReLU type layers on the training performance.

3.1. Training parameters tunning

Using the built in Matlab tool, Experiment Manager, we have tested the influence of the learning rate, the mini-batch size, and the validation frequency on the achieved accuracy. Table 3 presents the results obtained on the dataset containing spectrograms collected within the 2.4 GHz frequency range by modifying the initial training parameters presented in table 2.

The subtle differences in accuracy outcomes across different initial learning rate values can be explained by the ADAM solver's ability to adaptively adjust the learning rate value during the training process. Thus, a learning rate of 0.01 appears to be the most suitable choice as it strikes a balance between convergence speed and generalization capability, leading to better overall performance on training, validation, and test datasets.

The choice of minibatch size does not appear to substantially influence the accuracy of the model. This is explained by the fact that the minibatch size is closely tied to the processing capacity of the system. A larger minibatch size allows for more efficient processing by utilizing more data per iteration, potentially accelerating the training process while smaller minibatch sizes might necessitate more iterations to cover the entire dataset.

Increasing the validation frequency from 5 to 15 results in fewer validations performed during the training process. Consequently, this reduces the computational demand required for model evaluation. With fewer validations, the training process progresses more swiftly and leads to faster convergence and reduced training time. The improved computational efficiency comes with the cost of reduced accuracy for both training and test data that is not to be sacrificed when enough computational power and training time is available.

Table 3. Training parameter variation results

Training parameter	Value	Train accuracy	Validation accuracy	Test accuracy	Training time
LearnRate	0.001	100%	97.96%	97.25%	-
LearnRate	0.01	100%	98.75%	98.75%	43 min 25 sec-
LearnRate	0.1	100%	99.06%	97.4%	-
Minibatch size	16	99.29%	96.06%	97.61%	40 min 54 sec
Validationfrequency	15	96.38%	90.94%	94.06%	32 min 0 sec

3.2. Comparative analysis of activation functions

The selection of an appropriate activation function is very important in the design and optimization of neural networks, as it directly impacts the model's capacity to learn complex patterns and generalize to unseen data. A short description and theoretical comparation between the four types of activation functions (ReLU Layer, leaky ReLU, eluLayer, and swishLayer) chosen for this study is synthesized in table 4.

Table 4. Brief comparative analysis of the four activation layer types

Layer type	Description	Advantages	Disadvantages
ReLU	A simple yet widely used activation function that outputs the input directly if it is positive, and zero otherwise [16].	Efficient computation, encourages sparse activation, helps mitigate the vanishing gradient problem	Neurons may become inactive throughout training, resulting in gradients that cease to propagate effectively (dying ReLU issue).
leaky ReLU	A variant of ReLU that allows a small, non-zero gradient when the input is negative, preventing neurons from dying completely [17].	Addresses the dying ReLU problem, encourages non-zero gradients for negative inputs.	Adds a hyperparameter (the leakiness factor) to tune, which may require additional experimentation.
eluLayer	Another variant of ReLU that smoothly transitions to negative values for negative inputs, which can help alleviate the vanishing gradient problem.h	Smooth gradient everywhere, can capture negative values without a separate parameter like Leaky ReLU.	Slightly more computationally expensive compared to ReLU and Leaky ReLU.
swishLayer	Recently proposed activation function that introduces a non-linearity by multiplying the input by the sigmoid function of the input [17].	Smooth and differentiable everywhere, potentially better performance than ReLU and its variants in some scenarios.	Computationally more expensive than ReLU, may not always outperform ReLU in practice.

We have evaluated the performance of the four activation functions (layers) presented in table 4 on our dataset by modifying the network architecture presented in figure 4 and replacing the ReLU layer with the other activation function layers. Figure 7 presents the comparative evolution of the validation accuracy during training while table 5 presents the comparison between the four layers in terms of accuracy and training time on the combined 2.4 GHz and 5 GHz database.

Despite the inherent differences in their mathematical formulations and activation characteristics, one can observe that all four activation functions exhibit comparable results in guiding the model towards optimal performance. While Leaky ReLU demonstrates the highest validation accuracy the swish activation layer offers comparable performance with the advantage of faster convergence. One possible explanation for eluLayer's slightly lower performance could be its sensitivity to the dataset and network architecture. While the smooth transition of eluLayer may help in mitigating the vanishing gradient problem to some extent, it might also introduce subtle variations in the optimization landscape, affecting the model's ability to converge optimally. Additionally, the hyperparameter tuning required for eluLayer could also influence its performance.

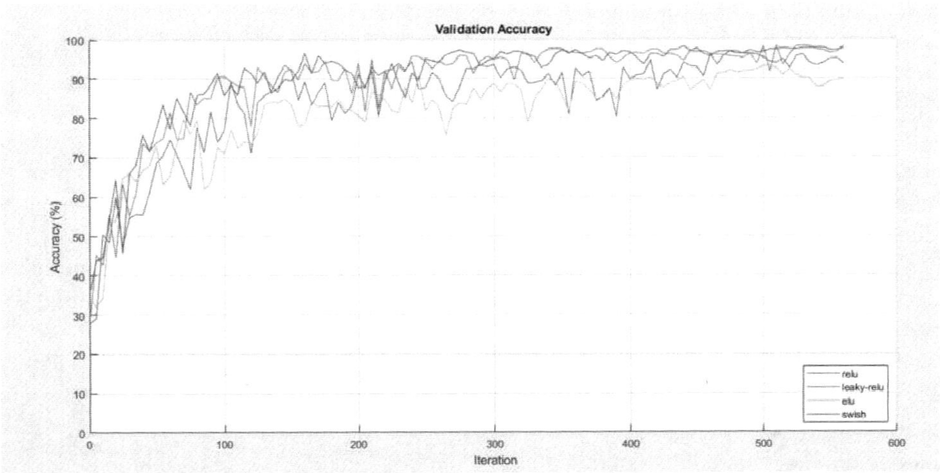

Figure 7 Comparative evolution of the validation accuracy during training

Upon comparison from the table, it's evident that leakyReluLayer exhibits the best performance. However, when considering the training time, ReluLayer emerges as the pragmatic choice, as it maintains respectable performance while being more computationally efficient. It's worth noting that the swish layer, despite offering comparable results, imposes higher computational complexity. Such insights are crucial for fine-tuning the network, allowing for informed decisions based on specific objectives and resource constraints.

Table 5. The accuracy and training time obtained for different activation layer types

Layer type	Train accuracy	Validation accuracy	Test accuracy	Training time
ReLU	99.50%	97.50%	97.50%	24 min 21 sec
leaky ReLU	99.96%	99.22%	98.75%	32 min 20 sec
eluLayer	98.39%	93.44%	95.94%	36 min 1 sec
swishLayer	99.42%	97.97%	97.50%	28 min 25 sec

4. Conclusions

In this study we have documented the construction of a spectrogram image database aimed at training a custom designed neural network to differentiate between six different Wi-Fi usage actions. Our study demonstrates the effectiveness of the proposed neural network powered by the ADAM solver in classifying Wi-Fi user activity from spectrogram images collected in real-world operating environment.

In addition, we have documented the impact of the training parameters such as learning rate, mini batch size and validation frequency on the model performance. Through empirical experimentation we have optimized the network performance to achieve a test accuracy as high as 98,75% by using a learning rate of 0.01 coupled with an appropriate validation frequency value (5).

The comparative analysis of activation functions reveals that while Leaky ReLU demonstrates the highest validation accuracy, the swish activation layer offers comparable performance with the added advantage of faster convergence. However, when considering practical constraints such as computational resources and training time, ReluLayer emerges as a pragmatic choice due to its respectable performance and computational efficiency.

The ability to classify Wi-Fi usage actions directly from spectrum can have direct applications in areas such as network optimization, security monitoring, and bandwidth management within wireless communication systems. By accurately identifying different usage patterns, organizations can enhance their signal intelligence capabilities, enabling better threat detection, network defense, and resource allocation.

References

[1] Daniele Atzeni, D. Bacciu, D. Mazzei and G. Prencipe. "A Systematic Review of Wi-Fi and Machine Learning Integration with Topic Modeling Techniques." Sensors (Basel, Switzerland), 22 (2022). https://doi.org/10.3390/s22134925.

[2] Chenshu Wu, Beibei Wang, O. Au and K. Liu. "Wi-Fi Can Do More: Toward Ubiquitous Wireless Sensing." IEEE Communications Standards Magazine, 6 (2022): 42-49. https://doi.org/10.1109/MCOMSTD.0001.2100111.

[3] D. Liu. "Research on Design and Development Trend of Wi-Fi Entertainment System of Civil Aircraft." 2022 4th International Conference on Applied Machine Learning (ICAML) (2022): 188-193. https://doi.org/10.1109/ICAML57167.2022.00043.

[4] Shayma Wail Nourildean and Abdulkreem Mohammed Salih. "Internet of Things based Wireless Sensor Network - WiFi Coexistence in Medical Applications." 2022 8th International Engineering Conference on Sustainable Technology and Development (IEC) (2022): 1-6. https://doi.org/10.1109/IEC54822.2022.9807574.

[5] Hosein, M., & Ramdass, K. (2022). Wi-fi Direct Applications within a Post Covid Classroom – Bridging the Gap between Fully Online and Face to Face Learning. 2022 International Conference on Information Networking (ICOIN), 464-469. https://doi.org/10.1109/ICOIN53446.2022.9687255.

[6] Apostolo, G. H., Bernardini, F., Magalhaes, L. C. S., & Muchaluat-Saade, D. C. (2021). A Unified Methodology to Predict Wi-Fi Network Usage in Smart Buildings. In IEEE Access (Vol. 9, pp. 11455–11469). Institute of Electrical and Electronics Engineers (IEEE). https://doi.org/10.1109/access.2020.3048891

[7] A. Thapaliya, S. Sengupta and J. Springer, "Wi-Fi spectrum usage analytics in University of Nevada, Reno," 2018 IEEE International Instrumentation and Measurement Technology Conference (I2MTC), Houston, TX, USA, 2018, pp. 1-6, doi: 10.1109/I2MTC.2018.8409865.

[8] T. J. Neal, D. L. Woodard and A. D. Striegel, "Mobile device application, Bluetooth, and Wi-Fi usage data as behavioral biometric traits," 2015 IEEE 7th International Conference on Biometrics Theory, Applications and Systems (BTAS), Arlington, VA, USA, 2015, pp. 1-6, doi: 10.1109/BTAS.2015.7358777.

[9] I.S. Faustov, V.A. Sladkikh, A.B. Tokarev and E.V. Koshcheev. "Detection and analysis of Wi-Fi signals for address direction finding." Radioengineering (2023). https://doi.org/10.18127/j00338486-202307-10.

[10] Hira Hameed, Lubna, Muhammad Usman, K. Arshad, Amir Hussain, Khaled Assaleh, Muhammad Imran, Q. Abbasi and James Watt. "Identification of Hearing Impaired People in Crowded Environments Using Wi-Fi Signals." 2023 IEEE International Symposium on Antennas and Propagation and USNC-URSI Radio Science Meeting (USNC-URSI) (2023): 279-280. https://doi.org/10.1109/USNC-URSI52151.2023.10237903.

[11] A. Azab and M. Khasawneh, "MSIC: Malware Spectrogram Image Classification," in IEEE Access, vol. 8, pp. 102007-102021, 2020, doi: 10.1109/ACCESS.2020.2999320

[12] Y. Zhang, B. Li, H. Fang and Q. Meng, "Spectrogram Transformers for Audio Classification," 2022 IEEE International Conference on Imaging Systems and Techniques (IST), Kaohsiung, Taiwan, 2022, pp. 1-6, doi: 10.1109/IST55454.2022.9827729.

[13] Safdar, Muhammad Farhan, Robert Marek Nowak, and Piotr Pałka. 2022. "A Denoising and Fourier Transformation-Based Spectrograms in ECG Classification Using Convolutional Neural Network" Sensors 22, no. 24: 9576. https://doi.org/10.3390/s22249576

[14] Miclaus, Simona, Delia B. Deaconescu, David Vatamanu, and Andreea M. Buda. 2023. "An Exposimetric Electromagnetic Comparison of Mobile Phone Emissions: 5G versus 4G Signals Analyses by Means of Statistics and Convolutional Neural Networks Classification" Technologies 11, no. 5: 113. https://doi.org/10.3390/technologies11050113
[15] Sârbu, A., Şorecău, M., Şorecău, E., & Bechet, P. (2023). Machine-Learning-Based Classification of Frequency Hopping in Radio Networks for Communication Reconnaissance. In International conference KNOWLEDGE-BASED ORGANIZATION (Vol. 29, Issue 3, pp. 78–84). Walter de Gruyter GmbH. https://doi.org/10.2478/kbo-2023-0079
[16] Abien Fred Agarap. "Deep Learning using Rectified Linear Units (ReLU)." ArXiv, abs/1803.08375 (2018).
[17] Yao Ying, Jianlin Su, Peng Shan, Ligang Miao, Xiaolian Wang and Silong Peng. "Rectified Exponential Units for Convolutional Neural Networks." IEEE Access, 7 (2019): 101633-101640. https://doi.org/10.1109/ACCESS.2019.2928442.

IOP Conf. Series: Materials Science and Engineering **1320** (2024) 012024 doi:10.1088/1757-899X/1320/1/012024

HF-NVIS Data Communication: OFDM Transmission Testing and Optimization Strategies

Emil Şorecău[1,2], Mirela Şorecău[1,2], Paul Bechet[2]

[1]Department of Communications, IT and Cyber Security Department, Sibiu, "Nicolae Balcescu" Land Forces Academy

[2]Department of Electrotechnics and Measurements, Technical University of Cluj Napoca, Cluj-Napoca, Romania

emil.sorecau@gmail.com

Abstract. Orthogonal frequency division multiplexing (OFDM), a technique that uses frequency division multiplexing to allow parallel transmission of data over multiple subchannels, is widely used in various wireless systems. Its effectiveness even extends to HF links, where it has demonstrated superior performance compared to single-carrier schemes. In addition, OFDM's robustness to multipath fading and its ability to efficiently utilize the available bandwidth make it a preferred choice for modern communication systems, but also for scenarios with complex communication links, such as HF range communications. The objective of this experimental research is to analyze and improve the performance of the OFDM technique adapted to HF communication channels. This study is motivated by the lack of a protocol to regulate HF communications using OFDM. The analysis and optimization method starts by considering the fundamental bandwidths of HF channels (ranging from 3 to 24 kHz). For this purpose, OFDM-specific modifications have been analyzed, including changes in the FFT size (representing the number of subcarriers), adjustment of the number and amplitude of pilots, analysis of the impact of packet length modification (pre-OFDM), and evaluation of guard intervals. In essence, our goal is to implement an OFDM technique commonly used in modern data communications in the context of traditional HF communications systems.

1. Introduction

High-frequency (3-30 MHz) link communications present significant challenges due to deep fading and inter-symbol interference caused by ionospheric propagation. HF channels are known to be variable, with rapidly changing propagation conditions and interference [1]. Multi-carrier modulation techniques, particularly orthogonal frequency division multiplexing (OFDM), offer a robust solution. OFDM splits data streams into orthogonal or nearly orthogonal sub-streams, improving subchannel transmission [2]. OFDM is widely used in today's wideband radio access systems due to its simple channel equalization, adaptable spectral resource allocation, and seamless integration with multi-antenna setups [3]. This versatility makes it a preferred choice for various wireless communication applications, including Wi-Fi, LTE and 5G networks [4]. However, ensuring reliable communication in wireless systems facing multipath propagation remains a significant challenge, mainly due to the critical need for accurate channel

IOP Conf. Series: Materials Science and Engineering **1320** (2024) 012024 doi:10.1088/1757-899X/1320/1/012024

estimation. In OFDM systems, interleaving pilot subcarriers with data symbols helps to track channel characteristics under varying conditions. However, fixed allocation of pilot subcarriers in OFDM systems can degrade performance when the channel impulse response changes rapidly [5]. This highlights the importance of using dynamic adaptation strategies to maintain system efficiency and reliability and to ensure seamless transmission at higher data rates, even for HF communication links.

HF communications face challenges such as signal degradation due to fading, multipath propagation and natural noise sources. Terrain and ionospheric conditions exacerbate these problems, making reception unpredictable. To overcome these challenges, one strategy is to split a single-carrier transmission into multiple narrowband channels. If the bandwidth on a single carrier is Fs, then by splitting we obtain narrowband subchannels, each of width Fs/N [6]. This approach simplifies equalization, reduces fading effects and improves signal reliability at different frequencies. The introduction of the FFT algorithm plays a key role in this process, simplifying signal analysis and allowing fast processing of multiple spectral components. The application of the FFT for discrete Fourier transform calculations streamlines multichannel communications, reduces the computational load, optimizes bandwidth utilization, and provides higher throughput rates with minimal inter-symbol interference [7].

OFDM is a highly effective data transmission technique for wireless broadband communications. It excels in facilitating rapid data transfers, even in scenarios with complex signal paths, where signals travel long distances and are reflected back by the ionosphere, such as HF communications [8]. OFDM splits a high-band signal into signals of lower frequency bands, optimizing the available bandwidth and effectively managing variable channel conditions. Its attractiveness lies in its straightforward approach to digital transmission. The efficiency of OFDM is attributed to the use of the Inverse Fast Fourier Transform (IFFT), which efficiently maps bits onto discrete carriers, providing performance simplicity that outperforms methods such as CDMA.

OFDM not only combats inter-symbol interference (ISI) and reduces inter-carrier interference (ICI), but also improves spectral efficiency by establishing orthogonality between subcarriers [9]. The basic principle is to divide the incoming data stream into N lower rate streams, each modulated on a distinct orthogonal carrier. This segmentation creates multiple subchannels within the channel, concentrating power into narrow frequency bands. The carriers are modulated independently, allowing individual optimization of the transmit power for each carrier. The independence of these subchannels ensures that interference between channels is eliminated at certain points, essentially converting the system to OFDM at the receiver. In traditional OFDM configurations, this crucial orthogonality between subcarriers is maintained using the Discrete Fourier Transform (DFT).

As mentioned earlier, OFDM allows data to be transmitted simultaneously over multiple subchannels via frequency division multiplexing. In coherent OFDM configurations, accurate channel estimation is essential to mitigate the effects of highly frequency-selective Rayleigh fading channels [10]. Channel estimation in coherent OFDM involves inserting pilot symbols at dispersed positions within the stochastic time-frequency signal, which produces noisy samples [11]. The design of channel estimators for OFDM requires an optimal distribution of pilot information in the time-frequency domain.

The effects of channel fading include distortion of both the amplitude and phase of the radio signal. In multi-carrier communication systems, each set of sub-carriers represents a different channel that experiences different degrees of fading [12]. To overcome this problem, specialists use certain subcarriers as pilots to gather information about the channel state. These pilot subcarriers are periodically inserted at known amplitude and phase intervals. The receiver uses these pilots to estimate channel state information and to interpolate channel states between pilots using linear or nonlinear techniques [13]. These estimates are essential for equalizing the carriers and rescaling their modulations for successful decoding. Despite the overlap in the frequency domain, the orthogonality of each subcarrier is preserved, reducing interference and improving overall system performance. In addition, advances in channel estimation techniques continue to play a key role in optimizing the efficiency and reliability of OFDM communication systems.

This paper extends the authors' methodology for designing, implementing, and testing a transceiver system using SDR platforms and GNU Radio/Python software applications [14, 15]. The focus is on the design of an RF communication system with NVIS propagation using the OFDM technique, which is widely used in modern data communications due to its advantages, and adapted to HF communication with limited channel bandwidths. The primary purpose of the proposed system is to perform packet error rate (PER) analysis by evaluating variations in bandwidth, FFT size, pilot number and amplitude adjustments, data packet length, and other relevant parameters. The performance of the HF-NVIS communication system has been tested under both laboratory and real-world conditions.

2. Materials and methods

The division of the frequency band into subcarriers allows the system to cope with channel impairments without the need for complex equalization filters, thus enhancing its overall performance. Overlapping subcarriers (Figure 1) is a distinctive attribute of Orthogonal Frequency Division Multiplexing (OFDM), wherein the range of individual subcarriers show mutual overlap. This phenomenon is made possible due to the orthogonality of the subcarriers, which enables them to coexist in close proximity without mutual interference. The amplitude peaks of one subcarrier align with the nulls of adjacent subcarriers, facilitating the compact arrangement of channels in the frequency spectrum. This architectural framework ensures the independence of demodulation for each subcarrier, despite their overlap.

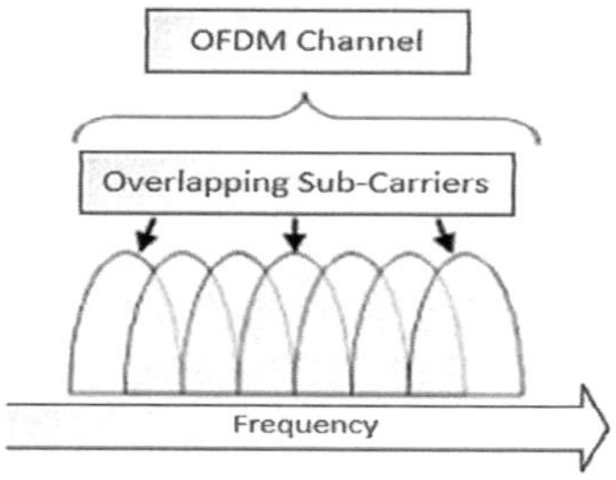

Figure 1. OFDM channel overlapping implementation.

Nevertheless, the deployment of OFDM is often challenging due to the necessity for precise synchronization. An effective approach to implementing and testing OFDM is to utilize software-defined radios (SDR) and GNU Radio. GNU Radio represents an open-source software tool designed to facilitate SDR systems implementation. In contrast to traditional hardware-based signal processing, this software-based approach allows for the rapid development and testing of complex software radio systems.

Figure 2 depicts the general framework of orthogonal frequency division multiplexing (OFDM), which was implemented in GNU Radio. It encompasses numerous subcarriers, represented by the vertical red lines situated at the middle of blue lines in the figure. These subcarriers comprise closely spaced frequencies, which are employed for the transmission of data. The term "guard subcarriers" pertains to the unoccupied spectrum present in an OFDM signal, with the objective of preventing interference with adjacent channels.

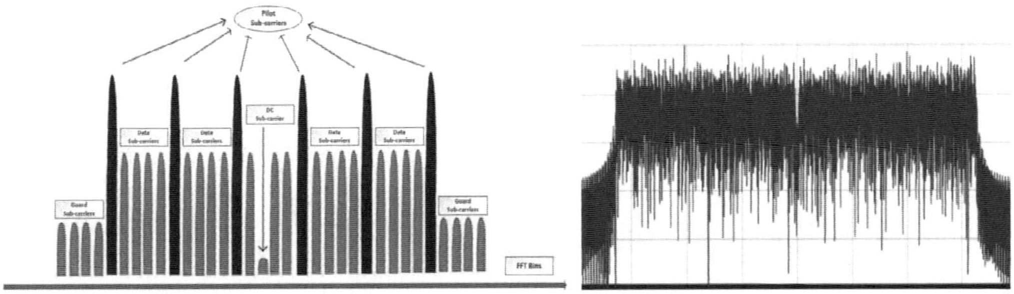

Figure 2. Standard sub-carriers allocation for an OFDM channel.

In an OFDM system, pilot subcarriers are specially placed among the data subcarriers and used for specific purposes, such as channel estimation to correct for distortion or fading affecting the signal during transmission. They provide a reference signal that can be used by the receiver to understand how the channel is affecting the data sub-carriers and to adjust the receiver accordingly. The FFT bins refer to the individual frequency slots within the OFDM signal, shown as the base of each subcarrier line. Each FFT bin contains information for a particular subcarrier, and the FFT is used to multiplex subcarriers at the transmitter and demultiplex them at the receiver.

Additionally, another component, the guard interval in the time domain, is the cyclic prefix (Figure 3). The cyclic prefix is a copy of the end of the OFDM symbol appended to the beginning of the OFDM symbol. It is used for two purposes. Primarily, it provides a period during which the receiver can synchronize without having to worry about moving from one symbol to the next. Secondly, the receiver receives a continuous signal representing the OFDM symbol, even if the symbol is temporally distorted due to the effects of multipath.

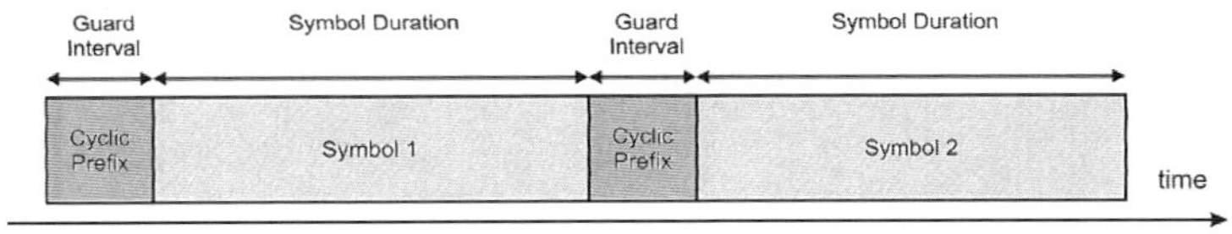

Figure 3. Guard interval implementation in time domain.

A collection of signal processing blocks is used to perform a variety of tasks in an OFDM implementation using GNU Radio. Some blocks can be seamlessly integrated into a signal flow graph and are provided as standard components within the GNU Radio library. Extensibility through the use of custom blocks written in Python is one of the strengths of GNU Radio. This makes it possible to create specialized functions which may not be included in the standard blockset. These custom blocks are then inserted into the GNU Radio flowchart like any other standard blocks, offering a more advanced solution while maintaining the general consistency of the system.

The implementation of the OFDM transceiver in GNU Radio is divided into two main software applications: transmitter and receiver, adapted to the HF-NVIS channel/link (Figure 4). The following presents the components of the system, both software and hardware:

- **Transmitter implementation**

The input data from a file is initially converted from a stream into a "labelled stream" at the beginning of the OFDM transmission chain. This involves converting the raw binary data into a format suitable for performing the modulation operation. The subsequent step is to incorporate a cyclic redundancy check (CRC) to identify any errors in the transmission. The data then undergoes a bit-packing process, in which one bit per byte is repackaged to align with the requirements of the modulation scheme. Digital modulation then transforms the bits into complex symbols, which is an essential step because it translates the binary data into symbols that can be transmitted over the radio spectrum. These symbols are then summed and the resulting modulated symbols are placed in the frequency domain at the appropriate carrier frequencies. Before transmission, a cyclic prefix is added to reduce interference between symbols caused by multipath propagation. Finally, the signal is resampled to a higher rate, interpolated, and filtered to prepare it for transmission over the wireless channel. This completes the transmitter function.

- **Receiver implementation**

Upon receipt of a signal, an OFDM receiver initiates processing by filtering the channel to reduce noise and interference. This is followed by a resampling operation, which reduces the sampling rate to match that of the original transmitted signal. This step is essential to ensure that the receiver is correctly synchronized with the transmitter.

The header and payload are then separated. The header is used for initial synchronization and signaling, while the payload contains the transmitted data itself. The signal is converted to the frequency domain, where channel equalization is performed to compensate for the effect of the channel on the signal. The signal is then converted back into its symbol sequences in the time domain. These sequences are demodulated into bit streams, which are reconditioned starting from 8 bits per byte, by reversing the transmitter conditioning. The CRC is checked to guarantee the integrity of the data received. Subsequently, the payload data is processed and written to a file, thereby completing the OFDM reception process.

- **HF-NVIS link**

In the HF-NVIS link, the channel itself is a central element, being integral yet independent of both the transmitter and the receiver. The OFDM signal is transmitted within this channel, and upon its passage through the wireless channel, the signal encounters a variety of distortions that can alter its integrity and performance. Among the distortions is noise, which is a significant form of distortion. It includes unwanted signals that are mixed with the original OFDM signal, which can degrade its clarity and accuracy. Similarly, interference, which can originate from other electronic devices or from overlapping frequency channels, can also disrupt the signal. Another significant challenge is fading, which refers to the variation in signal strength resulting from changes in channel conditions, such as atmospheric phenomena or multipath, where the signal travels through several paths to reach the receiver. To guarantee the quality and reliability of HF-NVIS communications, it is essential to effectively manage each of these distortions. In order to mitigate these issues and ensure robust communication even under adverse conditions, techniques such as adaptive equalization, error correction codes, and diversity schemes are frequently employed.

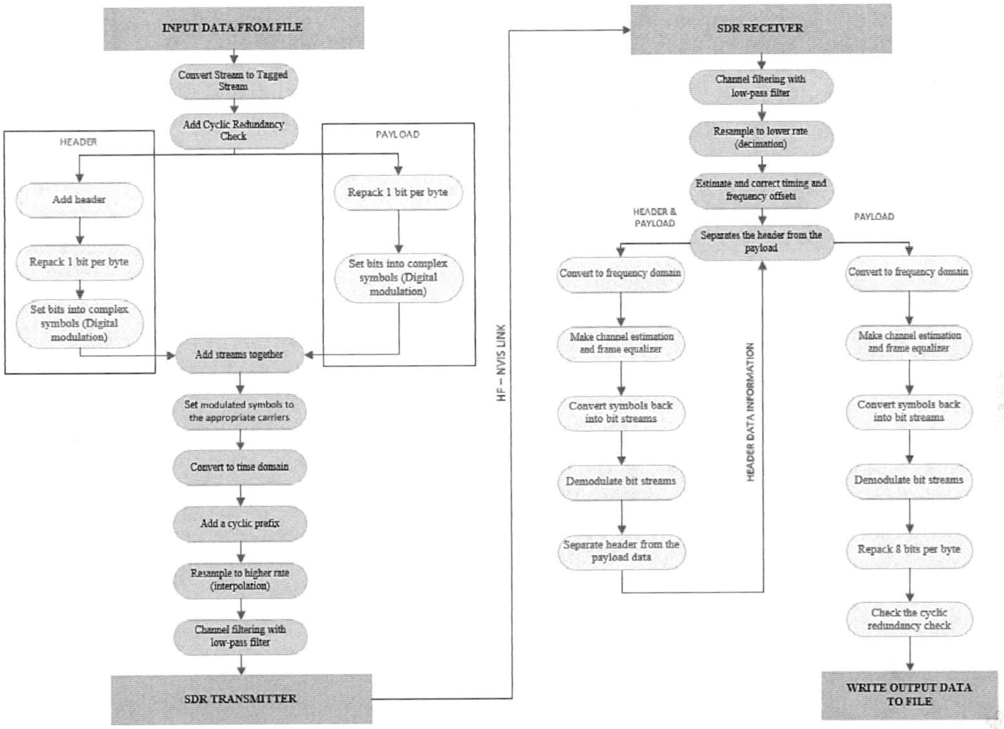

Figure 4. GNU Radio flowgraph design implementation for OFDM transceiver.

The implementation of HF transmitter and receiver configurations (Figure 5) was achieved through the utilization of the USRP N210 SDR. This was complemented by additional equipment, including a laptop with GNU Radio, low noise amplifiers (LNA), filters and power amplifiers, integrated hardware and software to facilitate efficient management of high frequency communications. Of these, the USRP N210 was identified as the most critical component, as it serves as the core of the system and provides a versatile platform for transmitting and receiving radio signals. The device is connected to a laptop with GNU Radio. The low-noise amplifier (LNA) enhances the received signal, minimizes noise, and optimizes the signal-to-noise ratio, which are crucial for HF applications. Moreover, band-specific filters are essential for isolating specific frequency bands and eliminating out-of-band interference, thus optimizing overall system performance. Finally, power amplifiers are integrated into the transmission line, thereby increasing the signal strength and allowing the transmitted signal to cover longer distances.

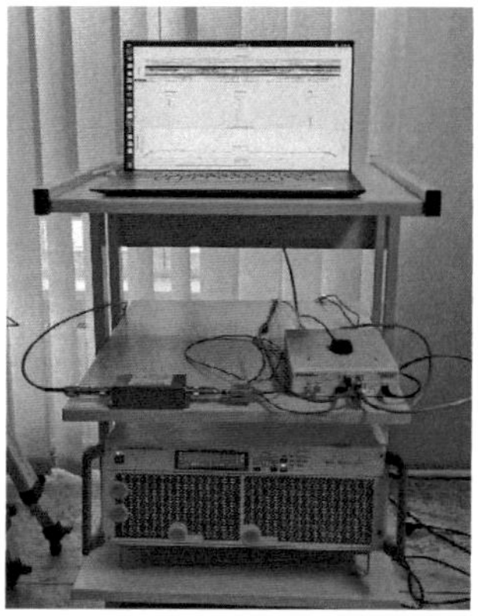

(a) **(b)**

Figure 5. Hardware system: HF-NVIS inside (a) and HF antenna "Inverted V" on a 10 m mast outside (b).

3. Results and discussion

The testing of the performance of an OFDM channel with regard to its packet error rate (PER) requires the variation of distinct variables. These include the number of subcarriers, the distribution of pilot signals, the amplitude of the pilot signals, the length of the cyclic prefix and the modulation techniques applied to the subcarriers. The table below presents a set of variables for the purpose of evaluating the performance of the OFDM communication system. The variables can be modified in order to ascertain their influence on PER, which serves as a metric for quantifying the frequency with which data packets are not received correctly.

Table 1. OFDM test settings.

Total number of carriers	Carriers allocated as pilot	Carriers without data	Length of the cyclic prefix	Digital modulation of carriers
32	(-7,-4,4,7)	12	5-20	BPSK, QPSK, 8PSK,
64	(-21,-7,7,21) (-23,-17,17,23) (-10,-4,4,10)	12	5-20	BPSK, QPSK, 8PSK,
128	(-51,-8,8,51,)	12	5-20	BPSK, QPSK, 8PSK,

The testing process enables a systematic variation of the parameters in question, while the measurement of PER allows the contribution of each factor to be determined. This understanding can be applied to optimize the OFDM system configuration to achieve a superior trade-off between data rate and reliability. The following results are presented for tests conducted according to changes in OFDM system parameters:

➢ **Test 1: Variation of cyclic prefix length**

The cyclic prefix length is the length of the buffer region located at the beginning of an OFDM symbol. This is used to protect against the effects of inter-symbol interference, which is a consequence of fading due to multipath propagation. Different lengths can be tested in order to identify the shortest usable prefix that provides sufficient protection without reducing data throughput.

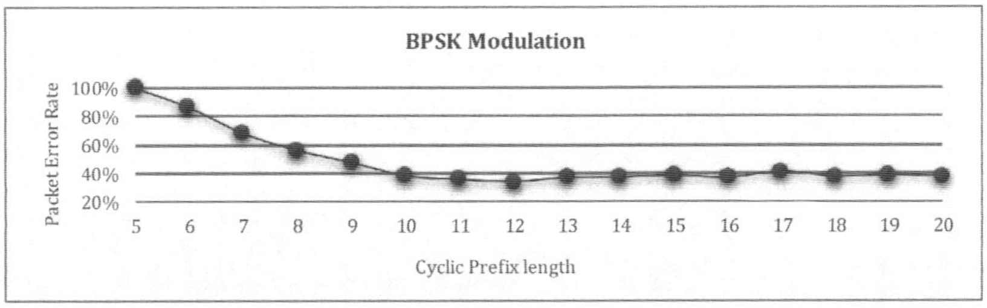

Figure 6. PER vs CP for BPSK OFDM modulation.

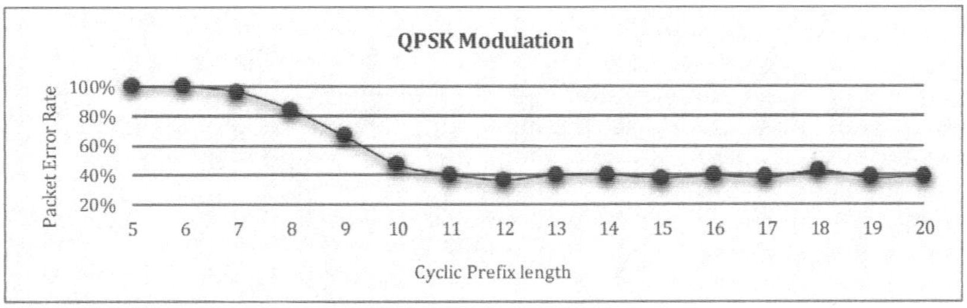

Figure 7. PER vs CP for QPSK OFDM modulation.

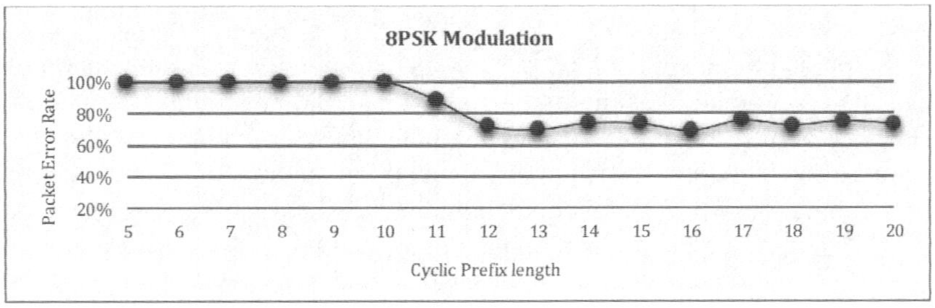

Figure 8. PER vs CP for 8PSK OFDM modulation.

The figures above illustrate the packet error rate (PER) as a function of cyclic prefix length for three distinct modulation schemes in an OFDM system: BPSK, QPSK and 8PSK.

In the case of BPSK modulation, the PER is approximately 100% when the CP length is 5. An increase in CP length to a value of 6 results in a notable PER reduction, indicating the necessity of a minimum CP length to achieve a lower error rate through BPSK modulation. While PER continues to decline with CP length, the rate of this decline gradually slows as CP length surpasses values 11 and 12, respectively.

The QPSK modulation exhibited comparable results to BPSK in terms of CP length. In contrast, for the 8PSK modulation, there was a discernible decrease in PER as CP length increased, only after the value of 10. The drop in PER remained relatively constant after the threshold of 12, as for the other two modulations.

In conclusion, the value of the cyclic prefix length must be adapted to both the propagation conditions and the type of modulation used, as the effect of the environment on the channel is different depending on the modulation.

- **Test 2: Distribution of pilot subcarriers**

The graph entitled "PER vs. Pilot Subcarrier Distribution" illustrates the relationship between PER and pilot carrier distribution in an OFDM system (Figure 9). The graph shows three different sets of pilot carriers, each color-coded. The orange color is used to represent pilot subcarriers at positions -21, -7, 7, and 21, the blue color is used to represent pilot subcarriers at positions -23, -17, 17, and 23, and the green color is used to represent pilot subcarriers at positions -10, -4, 4, and 10.

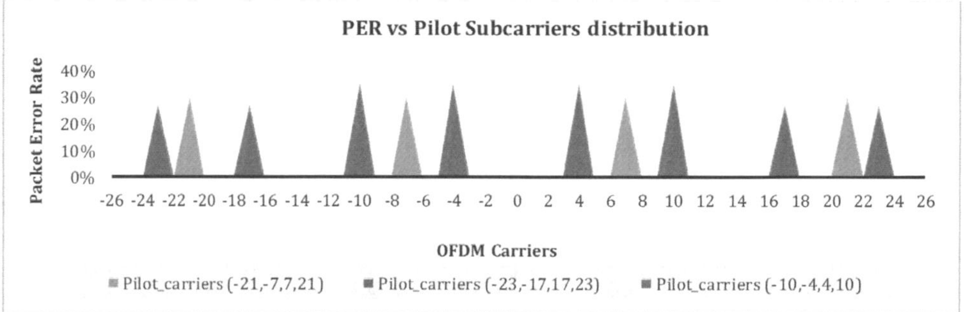

Figure 9. PER vs Pilot Subcarrier Distribution.

The first case, where the pilot subcarriers are arranged at positions -21, -7, 7, 21 (orange), characterizes a uniform distribution of pilot subcarriers across the OFDM spectrum. The second case, where the pilot subcarriers are -23, -17, 17, 23 (blue), expresses the situation where the four pilot subcarriers are placed at the two ends of the OFDM spectrum. The final case, where the pilot subcarriers are -10, -4, 4, 10 (green color), shows a higher concentration of pilot subcarriers around the middle of the frequency band. Upon analysis of the three cases, it can be observed that a wide uniform dispersion of the pilot subcarriers over the entire range of the OFDM signal facilitates a more accurate channel estimation, which is useful in multipath environments. This can be observed by the PER value, which is lower (case of pilot subcarrier distribution, orange color) and is specific to a channel that has a relatively uniform behavior in the frequency band.

However, there may be scenarios where the channel characteristics do not vary uniformly in the frequency band of the transmission channel. In these cases, a distribution of pilot subcarriers concentrated in the frequency areas of the transmission channel for which high variation is observed can be used. In such situations it may be useful to arrange the subcarriers unevenly in the transmission channel band. For example, the four pilot subcarriers in the two cases of the ionospheric channel can be arranged in the central area with green color or at the ends of the channel band with blue color.

In conclusion, the PER shows variations depending on the location of the pilot subcarriers. This demonstrates that the positioning of pilot carriers can have a significant impact on the error rate due to the way they are able to estimate the channel under varying conditions. From this perspective, the positioning of pilot subcarriers for the purpose of efficient channel estimation should be adaptive and lead to a reduction of errors at reception. Dynamic adaptation to the state of the ionospheric channel is a viable solution in this respect.

- **Test 3: Variation of the number of OFDM subcarriers**

The figure below depicts the correlation between PER and Fast Fourier Transform (FFT) size for three distinct modulation types. The FFT size, which corresponds to the number of subcarriers in an OFDM system, has a notable impact on the performance of the system, particularly in terms of its capacity to address interference and multipath effects.

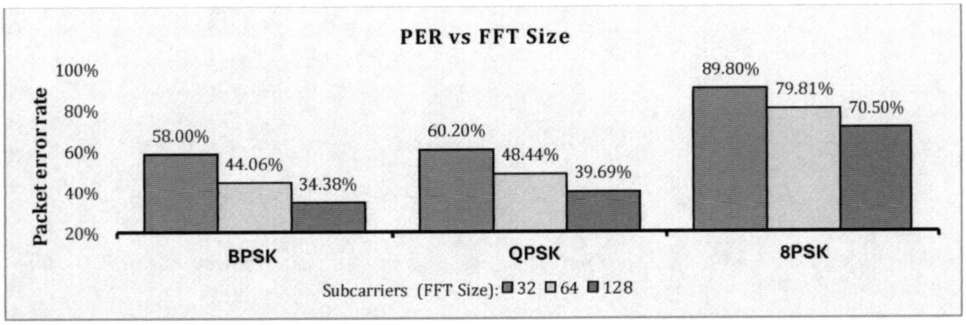

Figure 10. PER vs FFT size for different digital modulations.

The employment of 32 subcarriers results in the shortest FFT size with the highest error rates for all modulation types. In contrast, PER is reduced when the number of subcarriers is doubled from 32 to 64, regardless of the modulation type. This indicates that an increase in the number of subcarriers facilitates a more efficient distribution of the signal across the frequency spectrum, thereby reducing errors. Finally, for 128 subcarriers, the figure shows the lowest PER, indicating that further improvements may be expected due to better frequency resolution and more efficient use of available bandwidth. Nevertheless, the PER exhibits a slight increase in comparison to the aforementioned subcarrier configurations. This may be attributed to the elevated complexity of the system or the effects of densely packed subcarriers.

4. Conclusions

Software-Defined Radio platforms provide a versatile framework for evaluating and enhancing OFDM systems. Tests have demonstrated that variations in OFDM parameters, including modulation type, FFT size, cyclic prefix length, and pilot subcarrier positioning, have a significant impact on packet error rate. For instance, simpler modulation techniques such as BPSK exhibit lower Packet Error Rates compared to more complex schemes such as QPSK and 8PSK. Furthermore, it has been observed that increasing the FFT size improves system performance by reducing PER, although this also introduces more computational complexity and potential inter-carrier interference issues when arranging high-density subcarriers.

In the future, there is considerable scope for further optimization of OFDM systems by implementing adaptive modulation and coding schemes that dynamically adjust according to real-time channel conditions and traffic demands. This approach could significantly improve spectral efficiency and system robustness. Moreover, Moreover, further studies on the effects of pilot carrier distribution could result in the development of algorithms that optimize the placement of pilots based on prevailing environmental factors, thereby enhancing channel estimation and overall system throughput.

References

[1] Hill, E.E., "Frequency Agile OFDM System for Radio Communication in High Frequency Bands," *Electrical & Computer Engineering Theses & Dissertations,* 2022.

[2] Yli-Kaakinen, J., Renfors, M., Tuomivaara, H., "Multicarrier modulation for HF communications," *2016 International Conference on Military Communications and Information Systems (ICMCIS),* pp. 1-7, 2016.

[3] Bolcskei, H., "MIMO-OFDM wireless systems: basics, perspectives, and challenges," *IEEE Wireless Communications,* pp. 31-37, 2006.

[4] Saket, M., "Study of OFDM System for Application of Wireless Communication," *International Journal of Science and Research (IJSR),* vol. 6, no. 11, 2017.

[5] Majed, M.M., Omran, B.M., "Pilot Based Channel Estimation and Synchronization in OFDM System," *Journal of Engineering,* 2020.

[6] Pedersen, T.R., Reinisch, B.W., Paznukhov, V.V., Hamel, R., "HF propagation characteristics of artificial ionospheric layers," *XXXth URSI General Assembly and Scientific Symposium*, pp. 1-4, 2011.

[7] Tarniceriu, A., lordache, B., Spiridon, S., "An Analysis on Digital Modulation Techniques for Software Defined Radio Applications," *International Semiconductor Conference,* vol. 2, pp. 571-574, 2007.

[8] Kavitha, T., Sonika, M., Prathyusha, N.L., Raj, V.T., "Modelling of OFDM modulation technique in HF radio band using MATLAB," *Journal of Optical Communications,* 2022.

[9] Marchetti, N., Rahman, M.I., Kumar, S., Prasad, R., "OFDM: Principles and challenges," *New directions in wireless communications research, Springer,* pp. 29-62, 2009.

[10] Fernández-Getino García, M.J., Páez-Borrallo, J.M., Zazo, S., "Efficient pilot patterns for channel estimation in OFDM systems over HF channels," *Gateway to 21st Century Communications Village. VTC 1999-Fall. IEEE VTS 50th Vehicular Technology Conference,* vol. 4, pp. 2193-2197, 1999.

[11] García, M.J., Zazo, S., Páez-Borrallo, J.M., "Pilot patterns for channel estimation in OFDM," *Electronics Letters,* vol. 36, 2000.

[12] Sayed, A., Khatun, M., Ahmed, T., Piya, A.A., Chakraborty, P., Choudhury, T., "Performance Analysis of OFDM System on Multipath Fading and Inter Symbol Interference (ISI) Using AWGN," *Computational Intelligence in Pattern Recognition,* 2021.

[13] Saha, D., Grunwald, D., Sicker, D.C., "Wireless innovation through software radios," *Comput. Commun. Rev.,* vol. 39, pp. 62-67, 2008.

[14] Sorecau, E., Sorecau, M., Craiu, N., Sârbu, A., Bechet, P., "SNR Measurement of Ionospheric Channels for Availability Evaluation under NVIS Propagation," *International Conference and Exposition on Electrical And Power Engineering (EPE),* pp. 275-279, 2022.

[15] Sorecau, E., Sorecau, M., Bechet, P., "Adaptive HF radio channels based on signal-to-noise ratio using SDR technology," *10th International Conference on Modern Power Systems (MPS),* pp. 1-5, 2023.

AUTHOR INDEX

Institute of Physics
Dirac House, Temple Back
Bristol BS1 6BE UK

ISSN: 1757-8981
ISBN 979-8-3313-1006-6

9 798331 310066

CURRAN ASSOCIATES INC.
proceedings.com